AF775913

Contents

Editors' preface

Students at school or university, and others without a university training in biology, may have the opportunity and inclination to study local natural history but lack the knowledge to do so in a confident and productive way. The books in this series offer them the information and ideas needed to plan an investigation, and the practical guidance to carrry it out. They draw attention to regions on the frontiers of current knowledge where amateur studies have much to offer. We hope readers will derive as much satisfaction from their biological explorations as we have done.

An aphid colony is the scene of a lively drama, that runs day and night. A range of insects come to eat or parasitise the aphids or to drink their honeydew. Other Naturalists' Handbooks deal with the natural history of two major groups of aphid predators, hoverflies and ladybirds. This book complements them by focussing on the behavioural interactions among these and other visitors to a colony of aphids, interactions that offer limitless opportunities for observation and experimental investigation.

S.A.C.
R.H.L.D.
January 1987

Acknowledgements

I am very grateful to the editors, particularly S. A. Corbet, for their patience, help and encouragement during preparation of this book, and for their invaluable suggestions on an earlier draft. My wife, Joanne, gave particular help and gave assistance with the colour plates. Dr. Francis Gilbert and Mrs. Kathleen Davidson gave many helpful comments. I thank my typists, Mrs. Anne Rotheray and Mrs. Kathleen Davidson, for their patience and expertise.

Naturalists' Handbooks 11

Aphid predators

insects that eat greenfly

GRAHAM E. ROTHERAY
Royal Museum of Scotland,
Chambers Street, Edinburgh

with illustrations by
G. E. and J. C. Rotheray

Pelagic Publishing
www.pelagicpublishing.com

Published by **Pelagic Publishing**
20–22 Wenlock Road
London, N1 7GU
www.pelagicpublishing.com

Aphid predators
Naturalists' Handbooks 11

Series editors
S. A. Corbet and R. H. L. Disney

ISBN 978-1-78427-510-5 Pbk
ISBN 978-1-78427-516-7 Hbk

Digital reprint edition of ISBN 0-85546-270-1 (1989)
Text © Pelagic Publishing 2024
Illustrations © G. E. and J. C. Rotheray 1989

All rights reserved. Apart from short excerpts for use in research or for reviews, no part of this document may be printed or reproduced, stored in a retrieval system, or transmitted in any form or by any means, electronic, mechanical, photocopying, recording, now known or hereafter invented or otherwise without prior permission from the publisher.

British Library Cataloguing in Publication Data
A catalogue record for this book is available from the British Library

Dedication: to my parents, John and Margaret Rotheray

1 Introduction

A predator is an animal which kills and eats other animals. Predation is important in nature because it is one of the three basic ways in which energy passes between living organisms, the other two being herbivory (feeding on plants) and parasitism (feeding on, but not usually killing, the host). Predation is a common process; almost all animals have predators. For many, predation is a constant threat and life can be seen as a succession of attempts to escape attacks. Through the long interplay between predators and their prey a great diversity of prey-capture and anti-predator devices have evolved.

Part of the aim in studying predation is to understand what these devices are and how they function. Are they effective? How do predators capture and eat prey and why do they do it in the way they do? Can prey animals co-exist with their predators? If so, how do they avoid being eaten? What impact do predators have on the size and distribution of prey populations? Can predators regulate prey numbers? Is there anything in common between various predators, such as lions, sharks and ladybirds? Because aphids are easy to find and slow to move aphid colonies offer an excellent arena in which to observe predators in action.

Apart from these basic questions there is considerable interest in using predators as a natural means of controlling pests. This is biological control. Unlike chemical sprays, predators do not contaminate the environment, and if sufficiently effective, they can offer a better long term solution to most pest control problems. One of the earliest successes came in 1889 with the introduction to California of an Australian predator, the ladybird *Rodolia cardinalis*, which reduced the scale insect, *Icerya purchasi*, from a major pest on citrus trees to one of minor importance. Since then, predators have been used widely in biological control projects and many are being evaluated for use in the future. Biological evaluation is necessary because only by understanding the behaviour and ecology of potentially useful predators can the most effective species be chosen.

Aphid predators have often paved the way forward in research on predation. They can be reared and handled with ease, are common and can be observed in the field. They offer a rich variety of study topics and they behave well under laboratory conditions. Much is still unknown and puzzling about them. In this book you will find enough information to plan and carry out your own investigations on aphid predators. The book explains how predators and aphids can be found and identified. It also explains essential concepts involved in predator–prey behaviour and ecology. Finally, it provides an introduction to the literature on predation against which your own investigations should be planned.

Finding aphid colonies

Aphids, sometimes called greenfly or blackfly, belong to the insect order Hemiptera, the true bugs, which includes cicadas, froghoppers and scale insects. Hemipterans may be known by their mouthparts, which are modified to form a tube used for piercing and sucking (fig. 1). These range from the beak-like rostrum with which bed bugs suck blood to the long thread-like stylets of aphids. Aphids feed on plant sap, tapping the phloem tubes by inserting their mouthparts deep into a plant. Most aphids spend nearly all their time feeding and so are relatively immobile. Often many individuals feed close together forming densely packed aggregations or colonies. This habit makes them easier to find. Colonies (pl. 2) form underneath leaves, on stems, tree trunks, flowers or roots, and are particularly common on the growing tips of plants. Some species are hidden within leaf curls or galls. The aphid faunas of some common plants are described in chapter 4. If the foliage of these plants is carefully examined at the right time of year, aphid colonies should soon be found.

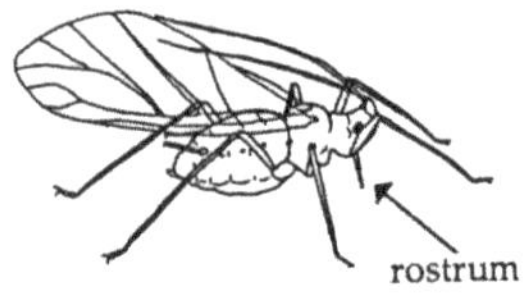

Fig. 1. Adult rose aphid *Macrosiphum rosae.*

Insects associated with aphid colonies

Many species of insect visit aphid colonies for a variety of purposes. A large group come to feed on the honeydew. Honeydew is a clear sticky fluid periodically ejected as droplets from the anus of a feeding aphid. It consists of excess sugars from the plant sap imbibed by the aphid and some waste products from its digestive system. Honeydew production can be observed easily by watching a feeding aphid for a few minutes. Often honeydew falls on leaves below the colony giving the leaves a wet, shiny appearance. When searching for aphids, this is a useful sign that they are close by. Possibly attracted by honeydew odour, many flies, bees, ants and wasps visit aphid colonies to feed on this sugary liquid. Throughout the day and night insects constantly arrive, feed and depart. Aphid colonies are probably just as attractive as flowers to many insects. Sooty moulds and fungi tend to grow in any uneaten honeydew, giving leaf surfaces a black, powdery look. This is another clue that aphids are close by.

Other visitors are aphid parasitoids. A parasitoid is an insect whose larva develops parasitically on or inside the body of its host insect. Unlike a true parasite, such as a liver fluke, a parasitoid normally consumes and kills its host. The adult parasitoids of aphids are black, wasp-like insects about the size of their hosts. Often the females are seen running excitedly around aphids tapping them with their antennae and probing them with the ovipositor or egg-laying tube. The parasitoids become still only when the ovipositor is pushed inside the aphid to deposit an egg. In the later stages of development, parasitised aphids can be recognised by their rounded, straw-coloured or dark appearance. After spending its larval and pupal period inside the aphid, the new parasitoid emerges by biting its way out leaving a characteristic hole in the dried husk of the aphid.

Sometimes a regular association will occur between an aphid colony and a nearby nest of ants. A worker ant will follow a trail from the nest to the colony, where she will shepherd the aphids, harvest their honeydew and return to the nest with her abdomen swollen with her load. Sometimes ants protect colonies by driving away predators and parasitoids and by carrying aphids to a new plant if the old one begins to wilt or die. There are circumstances, however, in which ants will give up their protective role and will eat aphids. Further details of ant – aphid relationships are discussed in chapter 2.

One of the largest groups of insect visitors tied to aphids are the predators. Aphid predators are of two types. Firstly there are the obligatory or aphid-specific predators, which depend on aphids for food. Included in this group are ladybirds, many hoverflies, aphid midges, lacewings, flower bugs and some solitary wasps and other flies. In many of these groups it is only the larval stages which eat aphids but in some, adults are predators too (see chapter 2). The second group are facultative or polyphagous (feeding on many different types of food) predators in which aphids are just part of a much wider diet. This group includes ground beetles, social wasps, birds, earwigs, and various groups of predatory flies. This book deals with examples from both groups.

Finding aphid predators

There are several methods by which aphid predators can be obtained for detailed study. When an aphid colony is found, carefully examine that part of the plant containing the aphids. Search next to and among the aphids. Avoid breathing on the aphids as they may fall off. Many predators are cryptically coloured (camouflaged) and difficult to spot, especially when they are young and small. Plate 1 shows the general appearance of the main groups of predators discussed in this book. On many plants, older stages of predators may hide in leaf curls or among flower petals, or move away from the colony when not feeding, so search these sites as well. Predators of many groups are nocturnal and a night-time search is necessary to find them. Before removing a predator into a tube or bag, see what it is doing. Most aphid predators are slow-moving, giving ample opportunity for observation of their behaviour.

Unless a long time is spent at each aphid colony these visual inspections and hand searches will not reveal every predator. A more effective method is to transfer the aphid-infested part of the plant into a plastic bag which is then sealed and left in a cool, shaded place (to reduce condensation). Usually within a day or so most of the predators can be seen on the sides of the bag. To ensure that you have them all, carefully search the plant remnant using a binocular microscope.

A third method of obtaining predators is to maintain short-term cultures. Eggs can be collected from aphid-infested plants or female predators can be induced to lay them (see p. 67 for techniques for incubating eggs and rearing larvae). For many experiments this is the preferred method, since it can supply large numbers of standardised individual predators which are uniform with respect to rearing conditions, age or size class.

These are the usual methods of obtaining predators. Other means include 'beating' the foliage of aphid-infested shrubs or trees. A white sheet, held taut on a wooden frame, is placed beneath a branch. The branch is then struck sharply several times causing the insects to fall on to the sheet. Obviously this method is inappropriate for delicate plants and predators which fly away rapidly! Leaf litter beneath aphid-infested plants can be searched at any time, because many predators end up there when they have finished feeding.

Another useful technique is to place laboratory-reared aphid colonies out in the field (Chandler, 1968a; Pollard, 1971*). Aphid-infested beans or nettles in plant pots can be used. These can be set out for a given length of time, a few days perhaps, and then searched in the laboratory for predators. Setting them at various heights and places can provide valuable evidence on predator habitat and oviposition site preferences and rate of colonisation.

* Publications cited in the text under the author's name are listed in Further Reading (p.72)

2 Natural history

Before collecting and experimenting with aphid predators, it is helpful to have a basic knowledge of the natural histories of the various groups. Only then can proper choices be made about which predators to work with; either because you find a particular group most interesting and challenging or because only certain predators are suitable for tackling a particular project. An outline natural history follows for each of the main groups of aphid predators.

Obligatory aphid predators

Hoverflies (Diptera: Syrphidae)

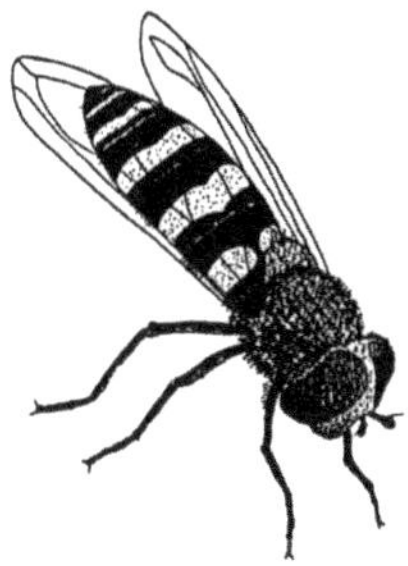

Fig. 2. A typical looking hoverfly adult – *Syrphus ribesii*.

Over 250 species of hoverfly (fig. 2) occur in the British Isles. About 100 of these have predaceous larvae and they belong mostly to the subfamily Syrphinae. In general, the rest either feed on plants (e.g. members of the tribe Cheilosiini) or on detritus in various situations such as mud and manure (e.g. Eristalini) or even in bee and wasp nests (e.g. Volucellini). The adult flies are not predators and feed on pollen, nectar and honeydew. They attract much attention because, unlike many flies, they are brightly coloured and are conspicuous when feeding at flowers or hovering in gardens and woods. They are an important group of insects in the pollination of flowers. In this respect it is interesting to note that the colour patterns of many species closely mimic those of bees and wasps. The book in this series by Gilbert (1986) gives further details of the adult biology and Stubbs & Falk (1983) give illustrated keys to adults.

Among the two-winged flies, the Diptera, predaceous hoverfly larvae are unique because of their diverse and contrasting colour patterns. In general, other Diptera larvae lack patterning and are uniformly coloured. Hoverfly larvae can be translucent with black and white or orange markings (*Syrphus*, *Episyrphus*); others are green with white stripes (*Epistrophe*, *Sphaerophoria*); and some are brown or mottled grey and black (*Metasyrphus*, *Dasysyrphus*). One species, *Meligramma triangulifera*, is apparently a bird dropping mimic (Rotheray, 1986). Probably these colours developed in response to a need to escape detection from visually hunting predators such as insectivorous birds.

Hoverfly larvae are cylindrical and they taper towards the head. They have no legs or eyes (plate 1.1). There are three larval stages separated by two moults in which the whole skin, including mouthparts, is shed. About half the total larval life is spent as a third-stage larva which eventually reaches a length of 8–17mm. At the end of the body is a tawny-coloured structure, the posterior respiratory process. The larva's respiratory system opens to the surface here. (A pair of much smaller openings occurs near the head.) This complex structure contains many morphological characters useful in distinguishing the various species. It consists of two plates which are separate in the first and second larval stages but fused in the third. At the front of the body the black arrow-shaped mouthparts are easily seen through the cuticle or outer skin.

Fig. 3. *Syrphus ribesii.* 3rd stage larva feeding on a sycamore aphid.

Hoverfly larvae are sucking predators; they pierce an aphid and imbibe the contents (fig. 3). During feeding, forward and backward movements of the mouthparts are seen helping to break up the tissues of the prey prior to swallowing. Whether digestive enzymes are injected into an aphid during feeding is unknown. When larvae are eating aphids, they often lift them from the plant surface.

Usually after finding an aphid colony, the gravid female lays one or more eggs close to the aphids. The whitish eggs are stuck to the plant and are about 1mm long. Chandler (1968b) gives an identification key to hoverfly eggs, although closely related species may be difficult to separate (Pollard, 1971). Within 2–3 days the first-stage larva emerges and begins immediately to search for aphids. Like many first-stage predators it will probably encounter several aphids before it captures one successfully. If it is a large aphid, the predator's first meal can take several hours, whereas a third (final) stage larva can consume such an aphid in a few seconds. Towards the end of the third stage, after having fed on anything up to 200 aphids or more, the larva stops feeding and the hind gut begins to empty. All through the larval growth period black waste products have accumulated in the hind gut. This can be seen by turning a larva over, or looking at the larva on a glass surface from beneath. The contents appear as a black smear on the plant. Any third-stage larva without this material in the hind gut has ended its feeding phase and is preparing to start the next developmental stage. Depending on the time of year, this could be aestivation (a prolonged resting period during summer), overwintering or pupation. Pupation occurs on the plant or in the leaf litter above the soil. Unless it overwinters or aestivates, the pupa gives rise to an adult within 9 days to 3 weeks.

Many hoverfly larvae feed at night, but occasionally starved individuals feed during daytime. Larvae up to about 7mm long spend the day resting in or near aphid colonies. Larger individuals tend to move away from the colony and rest in leaf curls and folds, or beside raised leaf veins and in similar hiding places. Leaf folds are useful places to sample, as often larvae of several species rest there together. Table 1 summarises some basic life history data for a range of common hoverfly species.

Hoverfly larvae are themselves attacked by a wide range of parasitoids belonging to the order Hymenoptera (which also includes the bees, ants and wasps). Almost every collection of larvae from the field will contain parasitised individuals. The commonest group are the Diplazontinae (Ichneumonidae) (see Fitton & Rotheray, 1982 for an identification key to genera and for biological notes; also Rotheray, 1979, 1981 and 1984). Other parasitoids include cynipoids and chalcids (see Scott, 1939; Rotheray, 1981). Parasitoids make a fascinating study in themselves and many of the techniques discussed in chapter 3 can be adapted to suit them.

Few quantitative assessments have been made to measure the impact of parasitoids on syrphid larvae. Next to nothing is known of how being parasitised affects predatory behaviour. For example, we do not know whether parasitised larvae consume more aphids or not. Even if there is no intention to study parasitoids, it is good scientific practice to keep records of any that emerge from hoverfly larvae (see chapter 5).

Table 1. *Life history data for some hoverflies*

Species	Larval feeding periods	Number of generations	Plants on which usually found	Overwintering stage
Syrphus ribesii	May–June July–Aug late Aug–Oct	2 or 3	trees, shrubs and low herbage	3rd-stage larva
Metasyrphus luniger	May–June July–Aug late Aug–Sept	2 or 3	low herbage	pupa
Platycheirus scutatus	May–June July–Aug late Aug–Sept	2 or 3	low herbage	3rd-stage larva
Episyrphus balteatus	(May–June) July–Sept	1 or 2 (partial spring generation)*	trees, shrubs and low herbage	adult
Epistrophe eligans	May–June	1	trees and other plants with aphids on leaves	3rd-stage larva
Sphaerophoria scripta	July–Sept	1 or 2	low herbage	3rd-stage larva

*In autumn most adult *E. balteatus* probably migrate south away from the British Isles so that the size of the spring generation depends on the numbers of adults that stay and overwinter successfully in this country.

Ladybirds (Coleoptera: Coccinellidae)

Forty-two species of ladybird occur in the British Isles. Most are aphid predators, but some feed on other small arthropods such as scale insects or mites rather than aphids. Some ladybirds are restricted to conifers, marshland or other special habitats and are not often seen (Pope, 1953; Majerus and Kearns, 1989). There are, however, a few ladybirds of widespread occurrence which are often recorded from aphid colonies. Even these ladybirds will eat other insects if they can catch them. Adults and larvae are predators of aphids but adults also take pollen, nectar and honeydew.

In contrast to the bright colours of the adults, ladybird larvae are slate-grey to black (pl. 1. 3). They have three pairs of legs. On the thorax and abdomen are rows of hairy, black, oval-shaped bumps called strumae. The colour, arrangement and type of bristles on the strumae are important aids to identification, as are the yellow, orange or white markings on the body, but these are variable and so not always reliable. At the rear of the larva is an anal clasper which helps it to move across leaves and stems without falling off. The larvae have biting and chewing mouthparts which are seen best from beneath. They can also inject digestive fluids into aphids and then suck out the partially-digested contents. Older stages tend to chew aphids rather than suck them.

Adult females lay batches of up to 40 yellow or white eggs, usually close to an aphid colony. After 2–3 days the eggs hatch and the new larvae feed on their egg shells. They may also devour any unhatched eggs. Some hours later the larvae begin their search for prey. There are four larval stages which can be distinguished on the basis of size. Fourth-stage larvae reach a length of 11–17mm. When growth is completed the larva attaches itself to the plant and the rounded pupa forms beneath the larval cuticle, which is then shrugged off. One or two weeks later the adults emerge. They are pale at first, but within a few hours they develop their adult colouration. Reversed colour varieties, with red spots on a black background, are known for many species (Majerus & Kearns, 1989).

Ladybirds spend the winter as adults. Large aggregations can occur within overwintering sites such as under bark or in leaf litter. Ladybirds will even enter houses to spend the winter in curtain folds or window frames. Most species have one generation a year with larvae present from May to July. Others such as the 14 spot (*Propylea 14-punctata*), the 10 spot (*Adalia 10-punctata*) and the 7 spot (*Coccinella 7-punctata*) may have two generations a year with larvae present from May to September.

A variety of parasitoids are associated with ladybirds. There are wasp-like braconids (Order Hymenoptera) belonging to the genera *Perilitus* and *Dinocampus* (Walker, 1961; Dean, 1983). There are also two groups of Diptera: small greyish and yellowish scuttle flies (Phoridae) in the genus *Phalacrotophora* which attack pupae, and housefly-like tachinids which have been reared from adult ladybirds (see Hodek, 1973; Majerus & Kearns, 1989).

Aphid midges (Cecidomyiidae: *Aphidoletes* and *Monobremia*)

There are four species of *Aphidoletes* and one of *Monobremia* in the British Isles. Of these, *Aphidoletes aphidomyza*, *Aphidoletes urticariae* and *Monobremia subterranea* are associated with aphids (Harris, 1973). The adult midges are very small (less than 4mm long) and delicate. They have long bead-like antennae. The larvae are orange and therefore very distinctive (pl. 1.4). There are three larval stages, the third of which grows to about 4mm long. If a third-stage larva is turned over under a lens or dissecting microscope, a small elongate structure, the sternal spatula, will be seen just behind the head. By arching its body round and using this structure, the larva can flick itself into the air if violently disturbed.

Fig. 4. Adult female *Aphidoletes aphidomyza*.

A. aphidomyza (fig. 4) is the best known of the three species. Females lay their eggs at night, singly or in batches of up to 40. These are orange and very small (0.3 x 0.1mm) but can be seen with the naked eye. The egg hatches after about 3 days. The first-stage larva attacks aphids at once, piercing the body and injecting a paralysing venom. Attacked aphids are immobilised rapidly and a first-stage larva takes several hours to suck out the body contents. Among larval predators *A. aphidomyza* is possibly unique in its use of a quick-acting paralysing venom (lacewing larvae may also inject venoms). The probable advantage is that the relatively large aphid is quickly overcome before it can start to defend itself. However, additional roles may be played by some components of the venom in the chemical breakdown of aphid tissues prior to ingestion. Larval development is completed in 7–14 days, after which aphid midge larvae enter the soil, where they spin a white papery cocoon in which to pupate. The adult emerges 1 to 3 weeks later.

To assess the importance of aphid midges as natural enemies of aphids, we need to know how many aphids may be immobilised and eaten during the total larval growth period. Published reports vary widely, ranging from 4 to 80 aphids per larva. Further study is required. It is possible that a larva may immobilise and kill many more aphids than it actually eats. For example Dunn (1949) reported that three

Aphidoletes larvae wiped out a 23cm-long colony of black bean aphids.

Larvae can be found at aphid colonies from about May to September and there are probably several overlapping generations each year. Aphid midges overwinter in the larval stage. It is not known definitely whether they feed at night or what their daily activities are. Aphid midge larvae have been recorded from aphid colonies in all kinds of situations – roots, trees and ground layer plants. Little is known of any parasitoids.

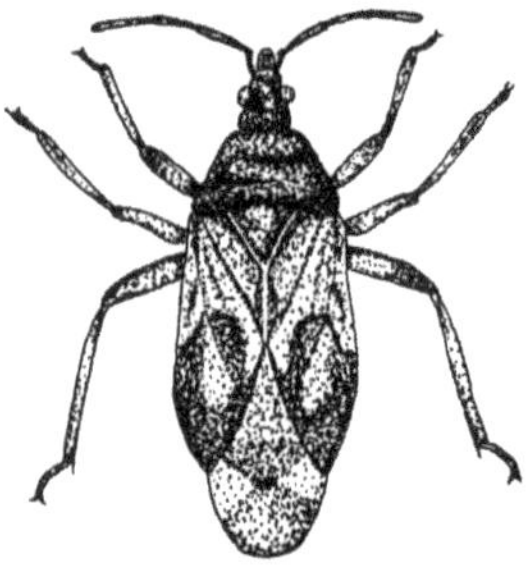

Fig. 5. Adult *Anthocoris nemorum*.

Flower bugs (Hemiptera: Anthocoridae)

The order Hemiptera includes many species which eat aphids, but the most important aphid predators are three species belonging to the genus *Anthocoris* (fig. 5). These are active, shining, red-brown to dark brown, beetle-like insects 3–5mm long (pl. 1.6). Like all bugs they possess a feeding tube or rostrum which is used to pierce and suck out the body contents of the prey. This structure distinguishes them from beetles, which have biting mouthparts. When a bug is feeding, its rostrum is extended ahead of its body. At other times, the rostrum is held up against the underside of the head and thorax, and can be seen by turning the insect over.

Both adults and larvae of *Anthocoris* eat aphids and probably any other live prey which they can capture. The larvae are similar to the adults but small and with undeveloped wings. The degree of development of the wing buds can be used to distinguish the five larval stages.

The eggs are laid singly or in small groups. They are inserted into plant tissue just beneath the epidermis which becomes raised up as a result and through which the exit cap or operculum protrudes (Sands, 1957). The first-stage larva emerges through the operculum when the egg hatches. Each of the five larval stages lasts 5 to 7 days. The fifth-stage larva moults to produce the adult which will mature for about 3 weeks before it is ready to mate. Flower bugs pass the winter as adults, under bark, in leaf litter and in similar sites. Most species have two generations a year, with larvae occurring first in May/June and then again in late July/August. A third late summer/autumn generation may occur also if conditions permit. Russel (1970) calculated that to complete development from the first stage through to the adult, *Anthocoris nemorum* needed about 50 second-stage sycamore aphids, whereas *Anthocoris confusus* required only 35 aphids. Anthocorids are recorded from a wide range of aphids on trees and ground layer plants. However, *A. confusus* feeds mainly on tree aphids (see Anderson, 1962). *Anthocoris gallarum-ulmi* larvae feed exclusively on *Schizoneura* aphids which cause leaf curls on elms (pl. 2.5) and *Anthocoris sarothamni* is also a specialist occurring only on broom (Southwood & Leston, 1959).

Lacewings (Neuroptera: Chrysopidae and Hemerobiidae)

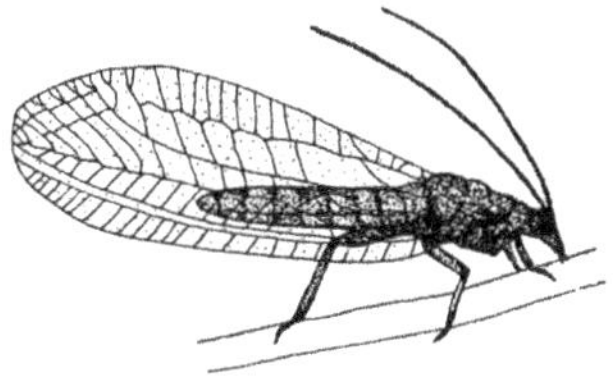

Fig. 6. Adult *Chrysoperla carnea* in typical lacewing 'resting' position with the wings folded over the abdomen.

Lacewings derive their English name from the complex pattern of veins in their wings (fig. 6). There are 54 species in the British Isles. They all have predatory larval stages, but some species have aquatic or semi-aquatic larvae (Sisyridae, Osmylidae). The non-aquatic lacewing larvae are aphid predators. Some adult Hemerobiidae will also take aphids. Chrysopid adults usually feed on honeydew and pollen but may take an occasional aphid (Fraser, 1959; New, 1975).

The larvae have three pairs of legs and look a little like ladybird larvae (pl. 1.2), but they are generally paler in colour, hairier and more elongate and delicate looking. Like ladybird larvae they have an anal clasper to help them grasp the plant. Their most distinctive feature is the brownish head capsule with its long curved mandibles. The mandibles are hollow. Lacewing larvae are sucking predators. They pierce an aphid with the tips of the mandibles and suck out the contents. They may inject digestive fluids and paralysing venom into the aphid prior to ingestion. Like hoverfly larvae, lacewing larvae often lift aphids when feeding. Some species, such as *Anisochrysa ventralis*, camouflage themselves by covering their backs with particles of debris, including prey remnants. Their abdomens possess specially hooked hairs to hold this debris in place.

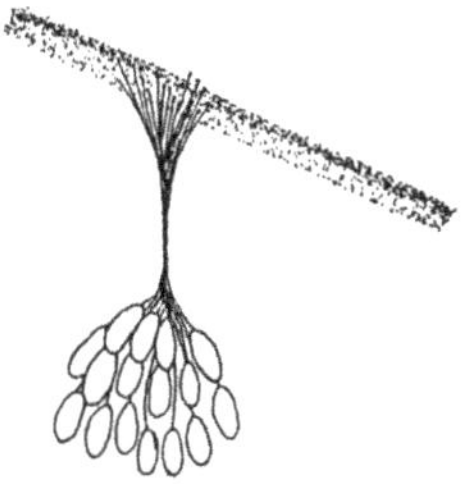

Fig. 7. Egg batch of *Nineta flava*, usually found under aphid infested leaves.

Some female lacewings lay their eggs directly on to the plant, as hoverflies do; others place each egg at the end of a long stalk. In some species, such as *Nineta flava*, females lay their stalked eggs in batches (fig. 7), while others, such as *Chrysoperla carnea*, spread them out. The eggs hatch in about 7 to 10 days. There are three larval stages over a period of 15 to 20 days. The third-stage larva spins a small, round, white cocoon within which to pupate. The adults emerge about 3 weeks later unless they are overwintering. Table 2 gives some information about the overwintering states and seasonal patterns of some common species.

Lacewings have a variety of hymenopterous parasitoids. The eggs may be attacked by tiny chalcid egg parasitoids (Killington, 1936) but apparently these have not been recorded yet in Britain. Larvae are attacked by *Charitopes* and *Dichrogaster* species (Ichneumonidae) and Anacharitinae (Cynipoidea) amongst others, but little is known of the ecology and behaviour of these parasitoids (see Killington, 1936; New, 1975; Canard and others, 1984).

Table 2. *Life history data for some lacewings*

Species	Egg	Larval feeding periods	Number of generations/ year	Plants on which usually found	Over-wintering stage
Hemerobius humulinus	stalkless, laid singly	May – Oct	2 or 3	deciduous trees and shrubs	prepupa in cocoon
Hemerobius lutescens	stalkless, laid singly	May – Oct	2 or 3	deciduous trees	prepupa in cocoon
Kimminsia subnebulosa	stalkless, laid singly	June – Oct	2 or 3	deciduous trees and low herbage	prepupa in cocoon
Eumicromus paganus	stalkless, laid singly	June – Sept	1 or 2	deciduous trees and shrubs	prepupa in cocoon(?)
Chrysoperla carnea	stalked, laid singly	May – Oct	2 or 3	trees, shrubs and low herbage	adult
Nineta flava	stalked, laid in batches	June – Oct	1 or 2	deciduous trees and shrubs	prepupa in cocoon
Nineta vittata	stalked, laid singly or in batches	June – Oct	1 or 2	deciduous trees	prepupa in cocoon
Chrysopa perla	stalked, laid singly or in small batches	May – Sept	1 or 2	deciduous trees, shrubs and low herbage	prepupa in cocoon
Chrysopa septempunctata	stalked, laid singly or in small batches	May – July Aug – Sept	2	deciduous trees, shrubs and low herbage	prepupa in cocoon
Anisochrysa ventralis	stalked, laid singly or in small batches	May – Sept	1 or 2	trees and shrubs	2nd or 3rd stage larva
Chrysotropia ciliata	stalked, laid singly or in small batches	June – July Aug – Sept	1 or 2	trees and shrubs	prepupa in cocoon

The above groups include the most common obligatory aphid predators. Details of other, less frequently encountered groups are given in table 3.

Table 3. *Infrequent or little-recorded obligatory predators of aphids*

Species	Notes	Further reading
Diptera:		
Chamaemyiidae *Leucopsis*	larval predators of aphids and their allies; larvae greyish; up to 6mm long; posterior spiracles on a prominent pair of tubercles; adults matt grey; 11 British species	Colyer & Hammond (1968)
Chloropidae *Thaumatomyia*	larval predators of root aphids; larvae translucent pale yellow, up to 6mm long; they move away from light; adults yellow and black; 5 British species, see Key I, couplet 13	Parker (1918)
Phoridae *Phora*	larval predators of root aphids; only one species has been studied; its larva is dirty white with hairy tubercles	Yarkulov (1972) – cited in Disney (1983)
Hymenoptera:		
Solitary wasps various genera e.g. *Passaloecus*	adults take paralysed aphids to nests in wood or soil and enclose several aphids in cells with an egg; the emerging wasp larva feeds on the paralysed aphids	Yeo & Corbet (1983)

Table 4. *Facultative predators of aphids*

Species	Notes	Further reading
Dermaptera:		
Earwigs	details in text	
Hemiptera:		
Damsel (Nabidae) and Capsid (Miridae) bugs	many species in these families occasionally feed on aphids	Southwood & Leston (1959), Alford (1984)
Coleoptera:		
Ground (Carabidae) and Rove (Staphylinidae) beetles	details in text	
Diptera:		
Doli's (Dolichopodidae) and Empids (Empididae)	large numbers of species in both families; adults are predators or carrion feeders on many insects; individual species can be as large as bluebottles or smaller than aphids	identification keys: Doli's – Fonseca (1978), Empids – Collin (1961); for a general account see Colyer & Hammond (1968)
Stilt legged flies (Micropezidae)	adults frequent shaded sites and feed on aphids and other insects; sometimes large numbers aggregate at individual aphid colonies; as their English name implies they have long legs; not all the 9 British species have been recorded feeding on aphids	Colyer & Hammond (1968)
Hymenoptera:		
Ants	details in text	
Social wasps (Vespidae)	frequent honeydew feeders and probably occasional aphid predators	Edwards (1980)
Arachnida:		
Spiders (various families)	few studies; they are probably more important predators than currently recognised	Jones (1983)
Harvestmen	few studies; nocturnal and can climb plants; unlike spiders they chew prey; remnants can be seen in gut analyses	Sankey & Savory (1974)

Facultative aphid predators

A large number of insects can be included in this group (see table 4). Their impact on aphids, however, is poorly known. This is due mostly to the unpredictable nature of their association with aphids. Their hunting and foraging behaviour has developed to cope with a wide prey spectrum, rather than just aphids. When hunting, they probably behave in an opportunistic manner and attack anything which is likely to be edible. They may not search specifically for aphids, nor stay close to them between feeding bouts. Unlike many obligatory predators, they can give up feeding on aphids and go on to something else quite easily. Facultative predators are, perhaps, more common at large aphid colonies or in places where several aphid colonies are close together, since such aggregations are more easily encountered. Nonetheless, in some well-studied aphids, such as those in cereal crops, facultative predators seem to be important (Edwards and others, 1979). In this book, earwigs, ground beetles, rove beetles and ants are discussed because, despite the above reservations, species in these groups feed quite commonly on aphids.

Earwigs (Dermaptera: Forficulidae)

It may seem a little surprising that earwigs are considered important aphid predators. Earwigs are familiar insects with their fearsome-looking forceps. Indeed the forceps are used in defence: if cornered, an earwig will raise up its abdomen in a scorpion-like fashion and open its mandibles wide ready to dash forward against the attacker. However, earwigs are quite harmless since neither forceps nor mandibles can penetrate human skin.

Earwigs have biting and chewing mouthparts and are thought to be omnivorous (including both animal and plant material in the diet). Aphids are, however, often eaten. Studies elsewhere in Europe have demonstrated that earwigs are important predators on fruit tree aphids, but in this country, earwigs are thought to be important predators on cereal aphids. The author has seen them frequently among aphids on a variety of trees and herbaceous plants.

Fig. 8. Tarsus of *Forficula auricularia.*

In Britain, there are four species of earwig but only one of these, *Forficula auricularia*, is widely distributed and common. This species is distinguished in the adult stage from the other species by having an expanded second tarsal segment in the foot (fig. 8), and wings which project beyond the wing cases (elytra). Females lay eggs in shallow pits in the soil in winter or spring. Unlike the majority of insects, the mother earwig gives maternal care to her young by standing guard over them and being aggressive towards intruders. There are four larval stages, which can be distinguished by the number of antennal segments: the first stage has 8, the second has 10, the third has 11, the fourth has 12 and adults have 14.

Earwigs are common throughout the spring and summer, but being nocturnal are not often seen. Aggregations of adults and larvae rest by day in leaf folds, under bark or stones or in similar places. The choice of a resting site may be influenced by the proximity of a food source. At night they move out of their resting sites to forage for food.

It is not understood how important aphids are in the diet of these insects and little is known of their predatory behaviour. For instance, how do they capture and eat aphids? Do adults and larvae vary in this respect? How many aphids are eaten? Do preferences exist for aphids over other types of food or for particular species of aphid? Do they require a mixed diet or are aphids alone sufficient for growth and development? There is plenty of scope for monitoring the movements of earwigs on aphid-infested plants and recording their predatory behaviour. By spot-marking the elytra with paint individual earwigs can be identified to see if they visit the same aphid colony on successive nights. Nothing is known about their possible interactions with other aphid predators. For example, what happens when a hungry earwig meets a hoverfly larva – does it eat it or ignore it?

Ground and Rove Beetles (Coleoptera: Carabidae and Staphylinidae)

A large number of species are in these two groups – about 350 ground beetles and almost 1000 rove beetles. However, only a small number feed on aphids. Ground beetles are easily recognised by their generally dark, shining appearance, ovoid form, prominent eyes, conspicuous biting mouthparts and long antennae. Rove beetles resemble earwigs in shape and appearance with their exposed abdomen and short wing cases. Rove beetles, however, have no forceps.

Rove and ground beetles usually live in the top few inches of soil, in leaf litter, decaying logs and other similar places. Most are predators of a wide range of insects and other organisms in these sites, although some feed on fungi or other plant material. Their larvae are also predatory but are rather poorly known and not often seen. They are whitish, have well-developed biting mouthparts and sometimes have tawny-coloured plates on the upper surface. Rove and ground beetles are active and fast-moving insects which depend on vision, touch and smell to find

Fig. 9. Ground beetle and aphid prey.

their food (fig. 9). All the aphid-feeding species are predominantly nocturnal (Vickerman & Sunderland, 1975). They spend the day resting in the soil and sometimes go quite deep. Luff (1978) records *Harpalus rufipes* 25 cm underground. They are common throughout most of the season.

Much of the published work on these predators has been done on those species attacking cereal aphids. Most of the evidence that aphids are part of the diet comes from examination of gut contents. When the gut of the insect has been removed, squashed between a slide and coverslip and examined with a binocular mocroscope, aphid siphunculi and tarsal claws have been found (Sunderland, 1975; Vickerman & Sunderland, 1975). There is evidence that some beetles climb plants at night, possibly just to locate aphids, but few observations of this have been made. Beetles may ascend plants for reasons other than prey location, for instance to take off for flight. On the soil beneath aphid colonies there are often many aphids which have become dislodged from the plant by wind and rain, or which have deliberately dropped to escape predators (see section 3.4). It is possible that many beetles feed only on such aphids and never climb the plant. Alternatively, contact with aphids, honeydew or cast aphid skins may provide just the stimulus to climb the nearest plant.

Much useful work can be done by making direct observations on aphid colonies at night and recording the behaviour of any beetles which are found. Obviously ground layer colonies, as opposed to those on shrubs and trees, are best suited for this type of work. How good are the beetles at climbing the plant and finding the aphid colony? Which species are involved? As with earwigs, spot marking individual beetles is possible at selected aphid colonies to find out whether repeat visits are made on successive nights. Detailed observations of predatory behaviour would be very revealing, particularly at aphid colonies on plants other than cereals, where little recording work has been done.

Ants (Hymenoptera: Formicidae)

There are 42 species of ant in the British Isles but only about nine are commonly encountered. Most species are omnivorous and combine predation with feeding on sugary plant fluids such as nectar, sap and juices from fruits and berries. In general, foraging for food is a group activity; rarely do individual ants leave the nest to search for food.

When hunting, an ant is likely to attack any small animal it encounters. Between some aphids and ants, however, special relationships have developed which do not

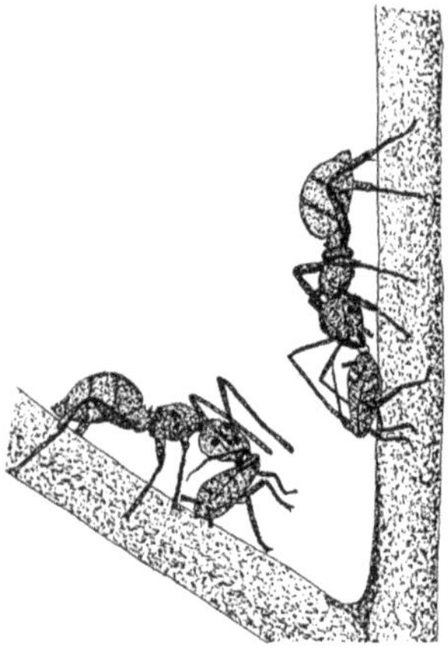

Fig. 10. *Lasius* ants feeding and carrying aphids.

involve predation. On encountering a colony of one of these species, instead of biting the aphids, the ant strokes them with its antennae. This causes the aphids to raise their abdomens and extrude droplets of honeydew which are sucked up promptly by the ant (fig. 10). Furthermore, ant-attended aphids change their behaviour. Instead of flicking and kicking honeydew away as usual, they retain it in their abdomens and wait to be stroked. If no stroking occurs they revert to their former methods of exuding honeydew.

Aphids which are associated with ants are termed myrmecophilous. However, many aphids have no regular relationships with ants, and in general these aphids have long siphunculi (used for defence, see section 3.4) and a long cauda (used to flick honeydew away).

At many aphid colonies, such as those on nettle, bean, rose or gooseberry, ants can be seen swarming over the aphids. In addition to taking honeydew, they keep the aphids close together and 'groom' the colony by removing cast aphid skins, wax droplets and other material. These relationships tend to be temporary, lasting at most a few weeks. More permanent relationships also occur between aphids and ants. *Forda* and *Trama* aphids are found almost exclusively inside ants' nests where they take their sap from grass roots and are closely attended by the ants. Additionally the ants will look after aphid eggs during the winter, and in the spring will place the newly-hatched individuals on nearby roots. Near the soil surface, ants sometimes build a protective collar of earth around the base of the stem where root aphids are present beneath. *Forda* and *Trama* aphids rarely kick their honeydew away and their siphunculi are short or absent. There is also a ring of hairs around the anus instead of a long cauda. All these adaptations aid the transfer of honeydew to the ants. A few aphids have developed further and show a kind of parasitism on ants. *Paracletus cimiciformis* aphids are found only in nests of the turf ant, *Tetramorium caespitum*, where they feed on regurgitated droplets produced by the ants. They rarely produce honeydew.

By attending aphids, ants not only gain a rich source of carbohydrates from honeydew but also obtain protein by eating the aphids themselves. The circumstances under which aphid predation occurs are not clear and much more work is required. Aphids which have withdrawn their stylets from the plant and are wandering around are more vulnerable to being eaten, perhaps because their walking movements elicit an attack response from the ants.

The situation can be more complex than this. Ants which are far away from their nests are less aggressive so that predation is more likely in colonies close to the nest (see

Way, 1963). Also, whether ants 'milk' the aphids for honeydew or eat them depends upon the general levels of food available in the ants' foraging areas. Way (1954) found that ants given a dish of honey gave up taking honeydew from coccids (close relatives of aphids) and ate the coccids instead. When the honey dish was removed, the ants reverted to their former behaviour. Pontin (1958) demonstrated that *Lasius* species regularly eat aphids of non-myrmecophilous species. The ant *Lasius flavus* will, however, feed on the myrmecophilous aphids in its nest.

Ants have other effects on aphids. El-Ziady & Kennedy (1956) demonstrated that *Lasius niger* ants attending the bean aphid, *Aphis fabae*, accelerated the rate of multiplication and growth of the colony and decreased the proportion of winged individuals among the adults. In addition, ants attending an aphid colony show 'ownership behaviour' – they act confidently and are aggressive towards intruders such as ladybirds and hoverfly larvae. After stroking ladybird and hoverfly larvae, *L. niger* ants picked them up and dropped them from the edge of a leaf or repeatedly attacked and drove them away (El-Ziady & Kennedy, 1956). Ladybirds encountered at a distance from the aphid colony were stroked but left alone. This protective function of ants against potential aphid predators requires additional study, the more so since contrary observations, that aphid predators are usually ignored, have been made (see Way, 1963). Perhaps when there are few colonies within the foraging area, ants protect their aphid colonies to a greater extent because they are a scarce resource.

El-Ziady & Kennedy (1956) used potted bean plants placed in a garden to record their observations on ant attendance and this easy method has much to recommend it, especially since ants are common in many gardens. Much basic information is needed. Is there any relationship between aphid colony size and the number of ants attending it? When does attendance start and end during the daily cycle? Many aphid predators are nocturnal – do ants attend aphids at night? What happens if the number of colonies is increased? Do ants select particular colonies to attend, such as those nearest the nest; or larger versus smaller colonies? Does their protective function increase or decrease with the numbers of colonies present or the distance from the nest? How do ants respond to a variety of aphid species presented to them? Are they selective, and if so, how and why? Under what circumstances are aphids eaten rather than 'milked' for honeydew? What, precisely, are the behavioural interactions between ants and other aphid predators?

3 Investigating predation

Anyone studying the biology of a species should begin by 'getting to know' the animal. Before starting a detailed investigation spend some time just watching your predators. Observe them closely. Study their morphology and see how they move and interact with one another. Sprinkle a few aphids in front of them and watch how the predators catch and eat them. See how they move on plants – do they manage equally well on flat leaves and narrow stems? How do they respond to bright sunlight? Are there any unusual habits or behaviour? Try to think down to their level and consider what problems they face. Each species has its own characteristics: you should be familiar with some of these before experiments begin. Indeed, much of the pleasure derived from biological studies is the specialised knowledge that comes from getting to know a particular group of animals very well indeed. Armed with a few hours' observation you will be in a much better position to design meaningful investigations and interpret their results.

Each of the following sections on what to eat, how to eat, searching behaviour and aphid defences serves as an introduction to a basic aspect of predation. Use this information as a springboard for your own studies. A first step could be to repeat some already published experiments using different species. Handling and other necessary techniques for setting up experiments are described in chapter 5.

3.1 What to eat: factors governing prey ranges

Over 500 aphid species occur in the British Isles in all sorts of situations and habitats. No one predator has a prey range covering all these aphids. Predators which feed on a single aphid species are described as monophagous but this situation is rare in aphid predators. Most predators are oligophagous, attacking a small number of aphid species. However, there are some very common predators which are polyphagous, accepting a wide range of aphids, for example, the ladybird *Coccinella 7-punctata*, the hoverfly *Syrphus ribesii*, the aphid midge *Aphidoletes aphidomyza* and the lacewing *Chrysoperla carnea*. For each predator we want to know which aphid species are eaten and why others are unsuitable.

Sometimes the distribution of adult predators gives a clue about the range of aphids eaten by the larvae; many lacewing species are known only from conifer woodlands, so aphids on these trees may be the larval food source.

One obvious method of investigating prey ranges is to sample widely for aphid colonies and record the predator species associated with them.Usually the presence of a predator within a colony is good circumstantial evidence that it can feed and develop on that aphid species. A great deal of this sort of sampling has already been done and the results have been used to make the general comments about prey ranges in chapter 2. Despite this, the prey ranges of many individual predator species are still poorly known. For instance, the predators of aphids in non-agricultural habitats such as hedgerows, road verges and waste ground are less well known, as are those predators of root and gall-forming aphids (see table 5). Furthermore the larvae of many hoverfly species are either unknown or poorly described. Detailed descriptions (for examples, see Dixon, 1960) with full biological data and a collection of specimens are needed before identification keys can be written. If you publish descriptions of new larvae, you should arrange to deposit the specimens in a museum, whose name should be noted in your paper, so that other workers can find and study the material easily in the future (see chapter 5).

Although widespread sampling of aphid colonies can give an impression of the total prey range of a predator, it does not reveal the whole story. Part of the difficulty is that predators, or females searching for colonies to lay their eggs, may only find the most abundant aphids, missing rarer but equally suitable species. Also, there may be suitable aphids which do not occur within the geographical region in which you are working. This does not mean that sampling is not worthwhile, only that for a fuller understanding, more work needs to be done.

Further clues about prey ranges can be found by studying the egg-laying behaviour of females and the developmental biology of larvae. If the factors governing egg-laying behaviour and larval development can be understood we can predict the probable prey range. For instance the syrphid *Sphaerophoria scripta* is usually most abundant in July and August and occurs in open sites (Pollard, 1971; Stubbs & Falk, 1983). Chandler (1968a) demonstrated experimentally that females prefer to lay their eggs on aphid colonies close to the ground. Therefore it can be surmised that the prey range of this species is probably limited to aphids in open sites, low down near the ground in July and August. A next step in understanding the prey range of *Sphaerophoria scripta* could be to study the developmental biology of its larvae fed on a number of aphid species from both inside and outside this region to test for their suitability in the manner outlined below.

Table 5. *Common root-dwelling, gall-forming and pseudogall (leaf-curling) aphids and their food plants*

Aphid species	Food plant	Notes
	Root-dwelling species	
Trama troglodytes *Protrama radicis*	various Compositae	large, pale, hairy aphids; *P. radicis* has conical siphunculi; *T. troglodytes* has no siphunculi
Rhopalosiphum insertum	various grasses (summer months only)	large, yellow-green aphids with short siphunculi
Aphis sambuci	umbellifers and docks (summer months only)	large, matt-black aphid
Anuraphis subterranea	umbellifers (summer months only)	small, shiny, black aphid
Anoecia corni (pl. 2.6)	various grasses	small, round, hairy, dark aphids with white markings
Forda formicaria	various grasses	round, creamy or greenish aphids often ant-attended
Smynthurodes betae	beans, brassicas and others	round, yellowish-white aphids
	Gall-forming species	
Adelgids	conifers	see Carter (1971)
Cryptosiphum artemisiae	mugwort	galls on leaves; whole leaf is curled over, bloated and turns red or yellow
Cryptomyzus ribis	red/blackcurrant, gooseberry	forms blister-galls on leaves
Schizoneura lanuginosa	elm	bloated galls on leaves at the tips of branches in the spring
Schizoneura ulmi (pl. 2.5)	elm	bloated, pale galls on leaves low down on the tree in the spring
	red/blackcurrant, gooseberry	on roots, summer months only
Pemphigus bursarius (pl. 2.3)	poplars	forms a smooth, round, green and red gall on the leaf in the spring
Pemphigus spirothecae	poplars	forms a spiral-shaped gall on the leaf petiole in the spring
Pemphigus filaginis	poplars	forms a blister-gall on the mid-rib of leaves in the spring
	Pseudogall (leaf-curling species)	
Phyllaphis fagi	beech	sometimes curls the leaves
Rhopalosiphum padi/ *Brachycaudus helichrysi*/ *Myzus cerasi*	cherry & other fruit trees	
Aphis rumicis	docks	
Aphis grossulariae	gooseberry	

Obviously predators must find their prey suitable; they must be able to capture, eat and grow on the aphids amongst which they find themselves. It may be that they can develop on a wider range of aphids than the female uses for egg laying, suggesting that oviposition is a limiting factor. Alternatively, larvae may not develop equally well on all the aphid species which females choose for egg laying. Blackman (1967) found that the 2-spot ladybird, *Adalia 2-punctata*, had a slower larval development, higher mortality and lower adult weight when fed on the bean aphid, *Aphis fabae*, than it did on the pea aphid, *Acyrthosiphum pisum*, whereas larvae of *Coccinella 7-punctata* did equally well on both prey species. Anderson (1962) showed that larval *Anthocoris confusus* could develop on *Aphis fabae* but no eggs were produced by the subsequent adults. How suitable do other predators find *Aphis fabae*?

The relative suitability of an aphid species as larval food for a given predator can be evaluated by comparative rearing experiments (see chapter 5). As a way of measuring the performance of predators on various aphids, Blackman (1967) recorded development rate (length of time in each stage); percentage mortality (number dying in each stage as a proportion of the total number at the beginning of the stage); larval or pupal size measured at various times during development; fecundity (number of viable eggs that each mated female produces in her lifetime) and adult life span.

If you have access to a balance which is accurate enough to weigh the dried remains of an aphid, you can estimate the weight gain of the predator as it grows and determine how much food is removed from individual aphids. This can give you very detailed data about suitability. Russel (1970) was able to compare the performances of *Anthocoris nemorum* and *Anthocoris confusus* feeding on the sycamore aphid, *Drepanosiphum platanoidis*, using this technique. By rearing the predators in closed plastic boxes with a moistened foam floor, he overcame the problem of the aphid remains drying out and giving misleading results when weighed. Each day aphids were weighed before and after presentation to the predators so that the amount of food consumed could be estimated. The predators were also weighed before and after feeding to compare growth rates. Russel found that both predators ate similar amounts of food, removing on average 75% of the live weight of each aphid attacked.

If you compare the range of aphids actually eaten by a particular predator species in the field with the range of aphids it can live on in the laboratory you begin to see how much the predator's larval diet is affected by the adult female's selection of a place to lay her eggs. How good are females at selecting the best aphids for their offspring? A predator which usually occurs on ground-layer herbaceous plants may prove able to survive on tree-dwelling aphids which it would never meet normally in nature. We know very little about this.

3.2 How to eat: factors governing feeding behaviour

Sometimes a predator kills more aphids than it eats. What is the minimum number of aphids that a predator needs to eat to complete development? How many aphids are actually killed? How does the number killed vary among different species of aphid? How do small predators cope with large prey and *vice versa*? Will a starved predator feed faster, or attack a wider range of prey species? At what time of day do predators feed? How do temperature and humidity affect this timing? Under what circumstances are predators cannibalistic? These and many other aspects of feeding behaviour are of special interest to those concerned with predators as natural enemies of aphids which are pests. Therefore detailed studies of particular species can be very useful.

Of particular value in understanding feeding behaviour are two easily-measured variables – handling time and capture efficiency. Handling time is the interval between initial capture of an aphid and the time when the predator discards the remains or resumes its search for more aphids. It can be measured with a stopwatch. Capture efficiency is a measure of a predator's success in capturing prey. The usual way to express this is to count the number of successful attempts at capturing prey per unit time and divide it by the total number of successful and unsuccessful attempts in the same period. For example, in one hour's observation you might record that a predator makes 50 attempts to capture prey, and 25 of these are successful. Thus the predator had a capture efficiency of 50% (25/50 x 100 = 50%). Clearly, we expect the most effective predators to have short handling times (they can deal with more prey in the available time) and high capture efficiencies.

Estimates of handling time and capture efficiency can be used to answer some of the questions in the first paragraph of this section. For every predator, handling times and capture efficiencies vary according to several factors (see table 6). By varying one of these factors under the same experimental conditions we have an excellent way of investigating predator performance. For example, we could investigate the effect of predator size by determining changes in handling times and capture efficiencies for a range of larval stages attacking similar sized prey. You might expect to find small predators taking longer to handle prey and having lower capture efficiencies. Do they? How does prey size affect handling time and capture efficiency?

Evans (1976a) showed that in *Anthocoris nemorum*, the larger the individual predator was in relation to its prey, *Acyrthosiphum pisum*, the greater was the capture efficiency. However, Rotheray (1983) found that in the syrphids *Melanostoma scalare* and *Syrphus ribesii*, small larvae did best on small aphids and large larvae did best on large aphids. In this latter study mechanical effects were important. Large predators could not pick up small prey because the prey became stuck at the surface of the characteristic drop of sticky saliva which these predators produce to catch prey. They were not drawn into it as larger aphids tended to be. Also large aphids were not so readily captured by small predators because these aphids were able to defend themselves (see section 3.4).

Table 6. *Factors influencing handling times and capture rates in aphid predators*

Factor	Possible effects
increasing hunger	increases handling times and capture rates; see section 3.2
relative size of the predator and prey	interaction variable; in some species the larger the predator in relation to prey the higher the capture rate and the lower the handling time, but not always; see section 3.2
density of prey	interaction variable; some predators respond positively to increasing density, others do not; see section 3.2
density of predators	decreases capture rates via 'interference' effects; see section 3.2
time of day	some predators are nocturnal and may refuse to feed during the day; see chapter 2
ability of prey to defend itself	effective defence is likely to decrease capture rates and increase handling times; see section 3.4
substrate quality	may increase or decrease capture rates and handling times, depending on the locomotory efficiency of the predator on a particular substrate; see section 3.3
toxicity of prey	decreases handling times and capture rates; see section 3.4
environmental conditions (e.g. temperature, humidity and light levels)	little understood; probably up to a limit, capture rates and handling times increase with temperature and humidity; high levels of light may deter predation

The effect of hunger on handling times has proved surprising. You might expect that very hungry predators would have short handling times as they feed rapidly and move on to the next aphid. But some starved predators have longer handling times than those which have not been so starved, at least in the first few aphids eaten. This has been recorded in the syrphid *Syrphus ribesii*. Third-stage larvae starved for 48 hours had higher handling times than those kept from food for 24 hours (Rotheray, 1983). Similar observations have been made on other aphid predators so this may well turn out to be a general phenomenon. The most likely explanation comes from the recognition that from a food point of view, aphids consist of fluids and tissues. The tissues comprise muscles and body organs and, being attached to the integument, are probably more difficult to extract from the aphid than body fluids which are easily sucked up. If a predator is not very hungry it may take only the easy-to-imbibe fluids from an aphid and leave the tissues behind. On the other hand, a starved predator in need of food probably spends time in removing the tissues as well as fluids, thereby extending handling time. A study of changes in handling times by predators starved to varying degrees and then fed aphids until satiation would help to extend our understanding of this phenomenon. Such an investigation could also be linked to changes in capture efficiencies to see if hungry predators are better at catching prey than less hungry ones.

A factor which attracts much interest is the effect of aphid density on predators. Density refers to numbers per unit area, such as numbers of aphids per leaf or square centimetre or length of row in a plot of plants. Aphids vary naturally in their densities on plants as they colonise, reproduce and eventually leave. Usually damage to plants is greatest when aphid densities are high. Therefore, from the point of view of controlling aphids, the response of predators to changes in prey density is of considerable interest.

Many aphid predators aggregate at sites where aphid densities are highest, perhaps because these aphid colonies are easier to find or because there is enough food to keep them from dispersing. Some predators react in a density-dependent manner; others do not. For instance some gravid hoverflies lay proportionately more eggs at higher densities of aphids, while related species behave in a density-independent manner (fig. 11).

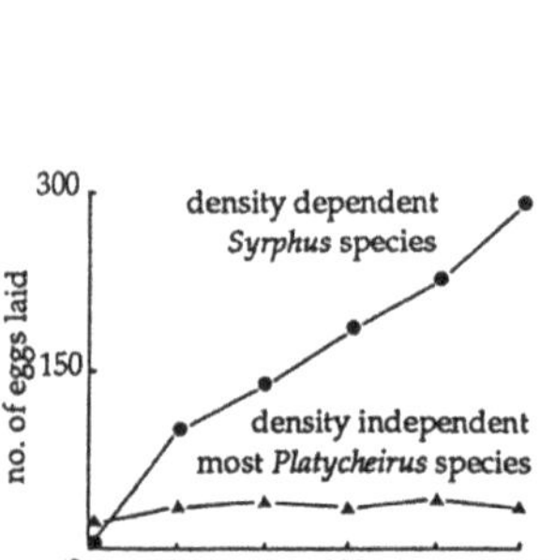

Fig. 11. Two patterns in egg-laying behaviour of hover flies in response to increasing size in aphid colonies.

A hungry predator's response to changes in prey density can be measured and can provide an excellent means of comparing different species of predator. The procedure is to expose individual predators to various

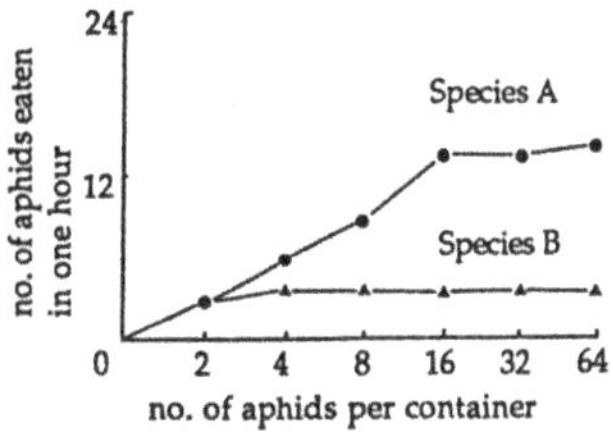

Fig. 12. Functional response curves for two hypothetical aphid predators, A and B.

densities of aphids for a given period and then record the numbers of prey eaten. For example, suppose we wish to determine the responses of two predator species, A and B, to varying densities of sycamore aphids. We would set up six containers for each predator species. Each container would have a similar-sized sycamore leaf in it. To create six different densities, 2, 4, 8, 16, 32 and 64 aphids would be separately placed on the leaves, one density level per container. After the aphids have settled, six hungry predators of each species are introduced, one to each container, and left for an hour. At the end of the period the predators are removed, the numbers of aphids consumed are counted and the data plotted graphically (fig. 12). The resulting plot is called a 'functional response curve' and it enables us to see at once how the predators have responded to density.

In species A we see that the response in the first part of the curve is almost linear. Later, at higher densities, the curve flattens out. In the second species B, there is little response to changes in density. From a practical point of view these results indicate that species A may be better at controlling high-density aphid colonies. However, we are also interested in why these differences exist and also why in species A the curve flattens out – why should the response not continue to be linear as at first? The answer to this latter question seems to be an effect of handling time. Most of the time is spent in capturing and eating prey at the higher aphid densities; little time is spent searching for them. In other words the predator is at the limit of its voracity for prey. The reasons why the two species differ in their response to density would need further study. Again it might be a question of handling time. Perhaps species B takes a long time to handle prey or requires fewer aphids to reach satiation. A thorough, comparative study of various predator species along these lines, combining functional response curves with behavioural analysis of the underlying mechanisms (see table 6), would be of considerable value.

The usefulness of functional response curves goes further than this. Not only can different predator species be compared but other variables affecting predator performance can be analysed also. For example, different stages of the same species can be compared by using them. The effect of hunger can be analysed by using predators starved for varying periods. The effects of prey species and size, temperature and humidity can be measured in relation to density and so on.

Allied to a prey-density response are effects of changes in predator density. In general it seems that the higher the density of predators the lower the feeding rate. In the few predators where this phenomenon has been investigated there is 'interference' between individual predators. Most commonly when one predator meets another they spend time interacting and may disperse from the site of the interaction rather than staying to feed. For instance, ladybird larvae occasionally fall to the ground after contact with eath other. Adult anthocorids may emit an alarm pheromone (a chemical messenger involved in communication between individuals of the same species) on contact with each other (Evans, 1976c).

The significance of interference is that it may cause predators to redistribute themselves among the available aphid colonies as they disperse from the site of interference (Hassell, 1976). From the point of view of controlling aphid pests, this could be an important mechanism, helping to ensure that a majority of aphid colonies is attacked.

Interference is likely between egg-laying females if they meet at the same aphid colony, but this is difficult to investigate because egg-laying females move fast and fly away. Investigations involving larval stages are easier. Ideally, experiments would be set up first to see if increased densities of predators depress the feeding rate, and second, to investigate the underlying behavioural mechanisms. Behavioural mechanisms are so poorly known that an investigation of these alone would be worthwhile.

To see if feeding rates are depressed, several containers, each with the same high density of aphids, are set up. Two equally hungry predators are placed in the first container, four in the second, six in the third, and so on. The predators are left for a given period. At the end of this time the predators are removed, the number of aphids consumed are counted and the results plotted graphically. A depression in the feeding rate will be revealed by a negative slope (fig. 13). By carefully observing the movements of individual predators and monitoring what happens when one predator meets another, clues about the underlying behavioural mechanisms can be gained. Are the predators aggressive towards each other? Do they disperse after the aggression ends? What effects do hunger level and predator size have on the response? Are alarm pheromones involved and what is their effect? Are the effects the same for all predator species – hoverflies, ladybirds, lacewings, and others? Are there aggressive, 'dominant' species which drive away other predators? Once again, there are ample opportunities for original and highly revealing observations to be made.

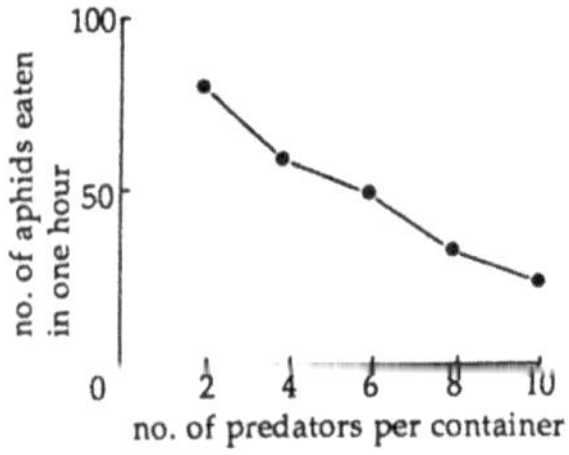

Fig. 13. Hypothetical effect of increasing the density of predators per container on the number of aphids consumed.

Sometimes physical interference between individuals of the same species results in cannibalism rather than dispersal. Cannibalism is known in most groups of obligatory predators. Chandler (1968) reports that he reared an *Episyrphus balteatus* larva (Syrphidae) entirely on other *E. balteatus* larvae. Under what circumstances does cannibalism occur? Often in cultures, large predators eat small predators. If aphid colonies are set up in containers, 'seeded' with various mixtures of species and stages and left for periods of a few days, cannibalism will probably occur. Which species are involved and how well do they survive? To what extent are facultative predators cannibalistic? How often does cannibalism occur in nature? When prey is scarce and predators are abundant, an advantage of cannibalism is that at least some individuals survive; in its absence perhaps none would.

3.3 Searching behaviour: finding prey on plants

How do aphid predators find their prey? If and when aphids are found do predators stay close to them and how is this achieved? Obviously a major part of the prey-finding process involves the egg-laying female. In her search for aphid colonies, a hoverfly uses scent, sight and tactile cues derived from the plant and the aphids themselves. Larval predators face a very different problem. With their lack of flight and poor sense of sight and smell they probably rely on tactile cues to recognise aphids. To find a source of food they must roam over plants and touch aphids. Yet plants are complex three-dimensional structures which change size and shape as they grow. Their surfaces can be hazardous to walk over being hairy or covered with slippery waxes and constantly mobile in the wind. Do predators use special searching mechanisms or is movement on plants entirely random? Do they search plants systematically leaf by leaf, or only particular sites on plants? What determines how long they spend searching in one site? If unsuccessful, do they try harder or leave for a new place?

Of course, for many larval predators aphids will be close by because the adult female lays her eggs next to them. However, larvae may face the problem of finding prey over greater distances, if, for example, they eat all the aphids in the colony before completing development, or if wind, rain or enemies knock them to the ground, or if aphids develop wings and fly away. Also some females lay their eggs well apart from an aphid colony. For example, hoverflies of the genus *Platycheirus* do this (Chandler, 1968c), and so do some ladybirds (Hodek, 1973).

To date, a number of searching mechanisms have been discovered which aid predators to find an aphid colony and keep close to it between feeding bouts. Bansch

(1966) reported that hungry aphid predators were positively phototactic (that is, they would move towards a light source) and/or negatively geotactic (they would move upwards, against the direction of gravity). These responses result in larvae moving to the tip of a plant. Since many aphids aggregate on growing tips this is clearly an advantage. However, not every growing tip has aphids, so what will predators do if they arrive at the top of a plant that has no aphids? Dixon (1959) showed that on narrow wooden cylinders (mimicking a plant stem) ladybird larvae eventually left the top because of a gradual change in the pattern of locomotion; the distance they walked between turns up and down the cylinder increased with time so that eventually they moved away from the top.

These general responses determine the overall direction in which searching predators move, but they do not govern the exact path followed. Banks (1957) observed the movements of individual lacewing and ladybird larvae on bean plants. He found no evidence of a systematic search. There seemed to be a considerable random component to their searching movements: larvae re-searched areas previously visited and spent time on the upper surfaces of leaves where aphids are rarely found. They also rested for long periods.

However, Banks (1957) noticed that his ladybirds spent a lot of time close to prominent leaf veins. This seems to be a common feature in aphid predators and as Dixon (1959) points out, it is alongside leaf veins that many aphids feed. This behaviour is clearly advantageous.

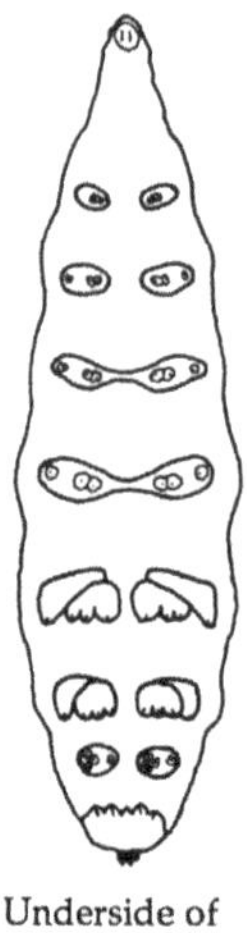

Fig. 14. Underside of *Metasyrphus luniger*, 3rd stage larva showing position of locomotory prominences.

Other surface properties of plants can be a serious hazard. For instance the waxiness, hairiness and smoothness of leaves hamper the movements of ladybirds (Shah, 1982; Banks, 1957; Carter and others, 1984). Indeed, falling from the plant is not a rare or insignificant event. 'Holding on' to plants has been a major hurdle which plant-dwelling insects have had to overcome (Southwood, 1973). Ladybird and lacewing larvae use their legs and a specialised organ, the anal clasper, to hold onto plants.

Hoverfly larvae, however, possess none of these structures. To hold on, they apparently use meniscus forces, created by coating the undersurface of the body with saliva (Bhatia, 1939; Roberts, 1971). During locomotion, larvae frequently leave a drop of saliva in front of them and move through it. In addition, many species have locomotory prominences on their undersurface. These cone-like structures occur in pairs on each abdominal segment. There is considerable variation in their structure. Those on larvae of *Epistrophe* and *Dasysyrphus* are flat and hardly raised at all, while in *Syrphus* and *Platycheirus* they are prominent and in *Scaeva* and *Metasyrphus* they are at their most complex. At the rear of these latter larvae the locomotory prominences are divided into four lobes each. On segments 5 and 6 they are bent over, so that their tips project backwards and, in conjunction with the anal lobe at the end of the body, form a 'U'-shaped grasping organ (fig. 14).

This organ is well suited to grasping stems, leaf petioles and leaf veins and it is on these parts of plants that these larvae are most frequently found. They move in a distinctive and characteristic way: the body is curled round the stem and the mouthparts and 'U'-shaped grasping organ moved alternately along the stem in a sideways motion (fig. 15). In contrast *Syrphus*, *Epistrophe* and other hoverfly larvae adopt the usual caterpillar-like looping movement on stems (fig. 16).

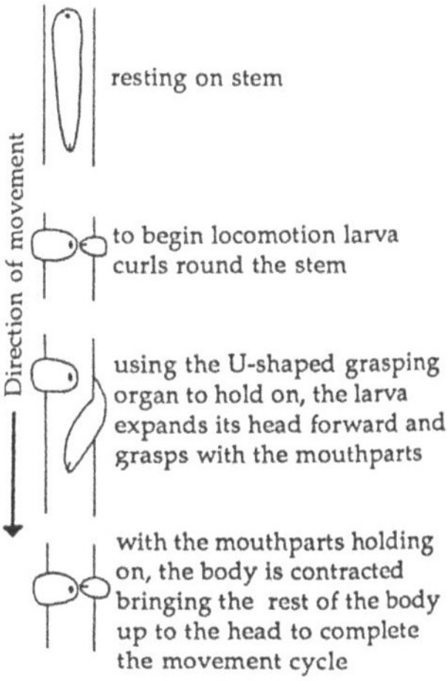

Fig. 15. Locomotion on stems in *Metasyrphus luniger*.

On flat substrates, such as underneath smooth leaves, where larvae of *Metasyrphus* and *Scaeva* cannot use their special grasping organ, they often fall off. It is on these substrates that *Epistrophe* larvae are better suited: they use all of their smooth undersurface to grip the plant and rarely fall off.

These differences in morphology and locomotion also affect the success of larvae in capturing aphids. On bramble stems infested with *Sitobion fragariae* aphids, I found the larvae of *Epistrophe eligans* had a capture efficiency of 13% as opposed to 68% on bramble leaves. On the other hand, the figures for larvae of *Metasyrphus luniger* were 24% on leaves and 78% on stems.

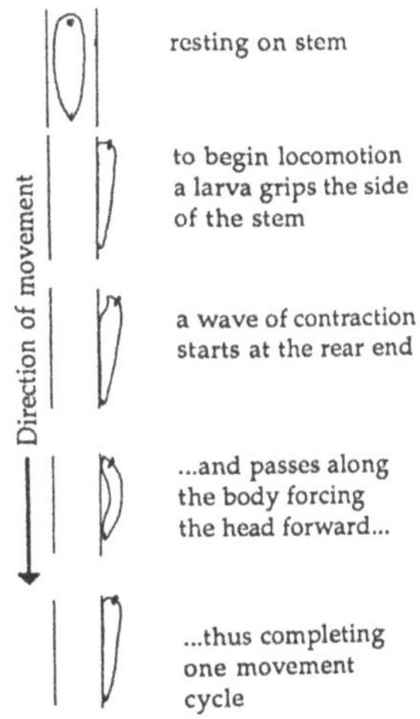

Fig. 16. Locomotion on stems in *Syrphus ribesii*.

There is still much to be discovered about plant structure and surface qualities as they affect locomotory efficiency. By careful observation of predator movements, it should be possible to identify what predators are most successful in locating aphids on a particular plant or plant structure. Obviously, such studies are relevant to the current research which is attempting to discover the best use of aphid predators in biological control.

There is also the question of predator hunting tactics. When a hungry predator encounters aphids it may begin 'area restricted searching' (fig. 17). This response involves a temporary increase in turning rates and a slower rate of forward locomotion. The result is an intensive search of a small area. 'Casting' movements may be involved. These are side to side movements of the head which increase the area scanned for prey. If no further encounters with prey occur, turning rates gradually decrease and speed picks up within a few minutes. If you present an aphid to a hungry larva this response will usually be seen after the larva has finished feeding. It has been recorded in anthocorid bugs (Evans, 1976b), ladybirds (Banks, 1957) and hoverflies (Chandler,

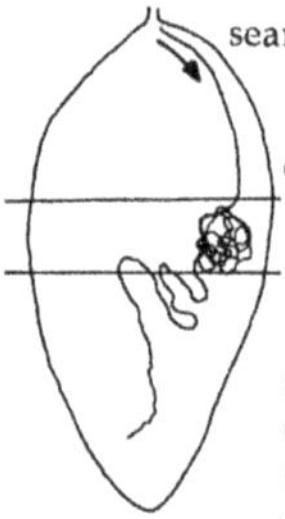

Fig. 17. Typical area restricted searching movements in an aphid predator on a bean leaf; black line indicates the course of movements.

1969; Rotheray & Martinat, 1984). It is of particular value to predators of aphids because, since these are colonial, contact with an aphid suggests that others are close by. Many of the factors which affect handling times and capture rates (table 6, p. 25) similarly affect area-restricted searching. These factors can be varied experimentally and their effects on turning or casting rates and speed can be measured, once again providing a means of comparing different species of predator.

Turning rates can be measured in the following way. The searching path of a predator is traced on to a sheet of glass or clear plastic held close to the predator, tangents are drawn to the path and a protractor is used to record the angles so formed. You can then add up the degrees of turning and express the results as a rate, for instance, degrees turned per minute. Speed is estimated by marking similar tracings of searching paths at appropriate intervals such as once every 5 or 10 seconds. The distances between marks are measured and the results expressed as centimetres covered per minute.

It is an obvious advantage for a predator to remain close to an aphid colony after it has finished feeding, since it will need several days or more to complete development. So what do predators do between feeding bouts? Banks (1957) found that satiated ladybird larvae stopped moving and rested in or near the aphid colony. In contrast, third-stage hoverfly larvae often move away from aphid colonies after feeding and hide in leaf curls, next to raised leaf veins, on bark or inside flowers (Rotheray, 1986). We do not know what other predators do.

3.4 Aphid defences against predators

Aphids are soft, colonial, and immobile for long periods and they give off enticing odours. All these properties make them attractive to predators. Yet aphids are far from being defenceless. Their habit of changing host plants at irregular intervals makes their occurrence notoriously unpredictable and threatens to leave predator larvae stranded, halfway through development.

Other more obvious ways in which aphids defend themselves involve the formation of protected or hidden places, such as galls. Some aphids cause the leaf on which they are feeding to curl, so covering them (table 5, p. 22). Other species live on plant roots. Here they may suffer less predation than aphids on the aerial parts of plants, but comparative studies are lacking. Some predators have specialised on root aphids (see table 4, p.14).

Other aphid species produce wax in the form of white, flakey material (the beech aphid, *Phyllaphis fagi*) or powder (cabbage aphid, *Brevicoryne brassicae*) which covers their bodies and surrounds the colony, making it difficult to pick out individual aphids. The effect of this material on predators is not understood. Are handling times or capture efficiencies affected? Do such aphids have fewer species of predator than those with no wax? Do predators have special ways of coping with the wax? *Meligramma cincta* is a syrphid particularly associated with *Phyllaphis fagi* aphids. Careful observation may reveal how this predator deals with the wax which its prey exudes.

Other aphids are known to be toxic. Vetch aphids, *Megoura viciae*, are regurgitated by larvae of the ladybird, *Adalia 2-punctata*, which may subsequently die (Dixon and others, 1965). The effect of this aphid on other groups of aphid predators should be studied.

On the exposed parts of plants, aphids have a variety of defence mechanisms involving behaviour. In many colonies aphids face down the plant or towards the leaf or stem base. Probably in these positions they can see an approaching predator and evade it. This is a topic to investigate. Compare how aphids manage to escape when approached from the side, front or from behind. How does the size of an approaching predator affect the ability of the aphid to escape? Can aphids move away from hoverflies and aphid midges as effectively as they can from ladybirds?

When a predator is close by, aphids can defend themselves by kicking, walking away or dropping from the plant. The propensity to kick, walk or drop varies from one aphid species to another but detailed comparative work is lacking (see Evans, 1976a). What happens to aphids which drop? How many find a food plant again? What determines whether they will kick, walk or drop?

If a predator grasps an aphid, another defence mechanism may be seen. The predator is [illegible] with a droplet of fluid from the siphunculi. In response to these fluids the predator may break off its attack and try to clean itself, which leaves the aphid free to escape. It would be valuable to have more information about this aphid behaviour. How successful is it in deterring various predators? How do predators cope with it? Which aphid species and stages produce these fluids – is it just those with long siphunculi?

In some aphids there is an 'alarm' pheromone within these fluids issuing from the siphunculi. It affects other aphids in the colony by making them withdraw their mouthparts and walk away or drop from the plant. In *Myzus persicae* the component considered responsible is trans-beta-farnesene. Such pheromones may help neighbouring aphids to evade their predators, but quantified data are lacking.

The pattern of dispersion can be protective. Colonies may be diffuse and spread over the plant, like the summer generations of *Aphis urticata* on nettle, so that they are difficult for predators to find. Aphids in other colonies, however, are close to each other. If in these colonies an individual begins to kick or walk away, its neighbours are disturbed and they, in turn, move off. In this way the whole colony becomes alerted. *Dactynotus* aphids on thistle stems provide an example.

When a predator approaches a colony the first aphids it encounters are those at the edge. These aphids may be at greater risk from predators than those at the centre of the colony. Such 'edge' effects have been recorded in the aphid parasitoid *Diaeretiella rapae* attacking *Brevicoryne brassicae* aphids on cabbage leaves (Akinlosotu, 1973, cited in Hassell, 1976), but what about predators? Some syrphid larvae seem to eat their way into colonies rather than concentrating on individuals at the edge, but again, little information is available. You can create circular colonies of aphids by attaching a cylinder of clear plastic to part of a leaf and carefully transferring aphids on to the leaf inside the cylinder until all the space is filled with aphids. Give them a short period to settle and feed, remove the plastic cylinder and you have a perfectly shaped 'colony' on which to test for edge effects.

Ants may protect aphids by driving away potential predators, removing debris from the colony and otherwise keeping it clean and free from fungal growth (see chapter 2 for more details).

Many of the components of feeding and searching behaviour described in this chapter have been incorporated into mathematical models of predation. Such models deal with very general properties of predation: they smooth over details and are simplistic. If they are to be useful tools in the management of aphid pest species, they must take account of the complexity of real systems. This requires a sound understanding at a basic level of the ecology and behaviour of predator – prey interactions. Attempts to construct realistic mathematical models have helped to show up areas which require further experimental study. Some of the investigations discussed in this chapter were stimulated in this way. Readers interested in pursuing this subject can find more in Hassell (1976) or Varley and others (1973).

4 Identification

Aphids

To identify aphids, collect a few adults from a colony. These will be the largest individuals. Winged specimens are always adults. There is usually only one species in each colony, but check other individuals for obvious differences in colour or shape and collect them if you suspect that another species may be present. Record the identity of the host plant. Plants can be identified using one of the many guides now available such as McClintock & Fitter (1982) or Phillips (1978).

The freshly-collected aphids should be placed in a drop or two of water and examined under a binocular microscope. Aphids can withstand short periods of submergence and the water helps to keep them still. Start by viewing the whole animal at x5 or x10 magnification to get an idea of its overall size, shape and colour. Orientate the aphid by moving the slide about or use forceps or pins to turn the aphid over and round. Increase the magnification for detailed views of the head and siphunculi which have many of the characters important for identification. Examine several individuals to get an impression of the range of variation present in your specimens. Using the guide (p. 37) turn to the section dealing with the plant species from which the aphids were collected and try to match the species description with the specimens you are examining. Most plants are attacked by several aphid species but usually only one or two are common. These more abundant aphids are marked with an asterisk. The guide describes the common aphid faunas of a range of plants. For help in identification of aphids on plants not included in the guide see Blackman (1974). Also see table 5 (p. 22) for gall-forming and root-dwelling aphids.

If your specimens do not genuinely match any of the descriptions then there are two possibilities. Firstly, the aphid may be one of the common, highly polyphagous (feeding on a wide range of plants) species. Try matching your specimens to the descriptions of these aphids given at the end of the guide. Secondly, it may belong to a group of rarer species not covered in this book. You can still work with unidentified aphids by giving them a reference number and storing a few adults in fluid (techniques, p. 71) for later identification. Further help in identifying aphids can be gained from Blackman (1974), which has coloured illustrations of some common species. The naming of rarer species not included in Blackman's book is a task for a specialist, and you could approach your local natural history museum, society or university for expert advice. Specialists will usually be pleased to help if your study is serious, but it is not fair to approach such people with many casually-collected specimens. Only aphids on plants in the guide can be identified in this book.

Predators

The first step is to find out to which taxonomic group the specimen belongs. Examine the specimen – at this stage it can be done by eye or hand lens as easily as with the microscope – and match it to one of the alternative statements in the first couplet given in Key I, p. 43. The number at the end of the statement will tell you which couplet to consult next. Continue until you reach a name.

Key I is a general key and will work for larval stages as well. It will permit identification of some of the common groups of visitors and facultative predators which can be found at aphid colonies. To take the identification of predators down to species, you need to check that the specimen possesses the features used in the keys. Check that its length is equal to, or exceeds, that given in table 7 for the groups where this is important. This precaution is necessary because many characters have not fully developed or are absent in very young stages. If your specimen is well below the length indicated then exercise caution is using the keys. If you rear it for a few days, all the necessary features should appear.

Table 7. *Minimum necessary predator length for Keys II to VI*

Predator group	Minimum length
Hoverflies, Syrphidae	6mm
Ladybirds, Coccinellidae	5mm
Aphid midges, Cecidomyiidae	3mm
Flower bugs, Anthocoridae	3mm
Lacewings, Neuroptera	5mm

Having identified the group, and, where necessary, checked the size of the specimen, use the appropriate key or guide to the major groups of predators. With a little familiarity it will soon be possible to go straight to these keys. For examination it is not necessary to kill the specimen. It can be placed in a drop of water on a slide or in

PLATE 1

1. *Syrphus ribesii*
   hoverfly
   3rd-stage larva (12 mm)

2. *Nineta flava*
   lacewing
   3rd-stage larva (10 mm)

3. *Adalia 2-punctata*
   ladybird
   4th-stage larva (6 mm)

4. *Aphidoletes aphidomyza*
   aphid midge
   3rd-stage larva (3 mm)

5. *Demetrias atricapillus*
   ground beetle
   adult (5 mm)

6. *Anthocoris nemorum*
   flower bug
   5th-stage larva (4mm)

7. *Tachyporus hypnorum*
   rove beetle
   adult (3.5 mm)

1

4

5

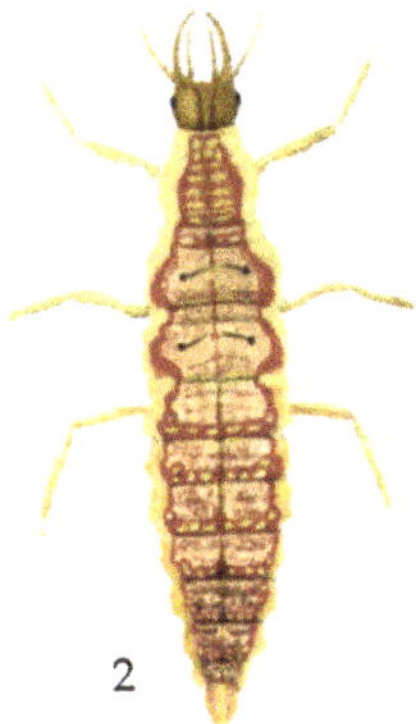

2

6

3

7

PLATE 2

1. *Dactynotus jaceae* (L.) stem-feeding aphids on knapweed (June–September) (*D. cirsii* and *D. aeneus* form similar colonies on thistles)

2. *Macrosiphum rosae* rose aphid on rose bud (May–July)

3. *Pemphigus bursarius* leaf-stalk gall of aphid on a poplar leaf (May–July)

4. *Hyalopterus pruni* leaf-feeding aphids on the underside of a plum leaf (May–June). This aphid migrates from fruit trees to reeds in summer

5. *Schizoneura ulmi* leaf-curling aphids on the underside of an elm leaf (May–June). Large colonies curl the leaf over completely to form a gall

6. *Anoecia corni* root-feeding aphids at the base of cocksfoot grass (June–August)

a small container. Very active individuals can be slowed down by placing them in a refrigerator for a few minutes. Do not use a freezer – this will kill them. To see the various characters, a binocular microscope with a range of magnification between x 5 and x 50 is the best instrument to use, but careful observation with a x 10 hand lens will also work in many cases. Killed or stored specimens can be identified with the keys but some groups lose their characteristic colours or shapes, so make notes or take photographs before storage.

In one or two places there is advice to rear specimens to the adult stage. This is because the larvae are so similar to those of another species that reliable distinguishing features have been found only in the adults. To identify these reared specimens, use the adult keys which follow the larval keys. These adult keys can also be used to verify larval identifications if subsequently you rear specimens through. This is a useful check and a good habit to adopt, especially when dealing with new or unfamiliar species. Sometimes, with the rush of fieldwork, there is not enough time to identify every specimen. Rearing them through, and identifying the adults later, is a convenient method of dealing with large samples.

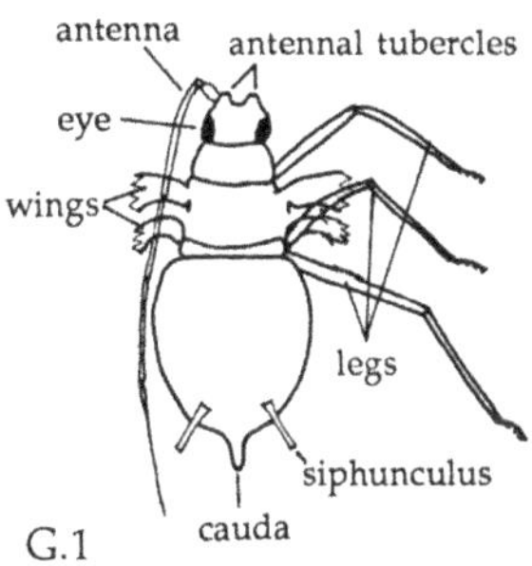

G.1

Aphid species guide

Broad beans or Field beans (*Vicia faba*)

Aphis fabae Scopoli*, the black bean aphid – A black or blackish-green aphid, often with white wax markings on the abdomen. They form densely clustered colonies on stems and the undersides of leaves.

Acyrthosiphum pisum (Harris), the pea aphid – Large pale green, narrow-bodied aphids with long siphunculi (G.1) and cauda (G.1). They tend to fall from the plant if disturbed.

G.2

Megoura viciae Buckton – A large, dark green aphid with a black head.

Aulacorthum solani (Kaltenbach) – A long-bodied greyish-green or pink aphid having long siphunculi covered at the tip with a network pattern of ridges (G.2).

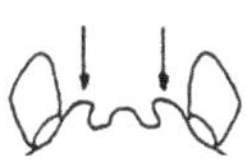

Birch (*Betula pubescens*)

Euceraphis punctipennis (Zetterstedt) – A light green aphid covered in a whitish waxy secretion. The adults live solitarily.

Betulaphis quadrituberculata (Kaltenbach) – A green aphid with rows of tubercles along the abdomen. Each tubercle ends in a tuft of hairs.

Glyphina betulae (L.) – A small, hairy, rounded, dark green aphid with white markings.

Bramble/Blackberry (*Rubus* species)

Sitobion fragariae (Walker)*, the blackberry aphid – A rounded, shiny green or pink aphid with greenish siphunculi covered in a network pattern of ridges. They are usually on stems but sometimes on leaves. They migrate to grasses in summer.

Amphorophora rubi (Kaltenbach) – A large, slender, shining yellow-green aphid with long legs and antennae and slightly swollen siphunculi (G.4).

G.4

Aphis ruborum (Borner) – A small, dark green aphid living on the undersides of the leaves.

Macrosiphum funestum (Macchiati) – A large green or reddish aphid. The long, tapering, black siphunculi distinguish it from the commoner *Sitobion fragariae.*

Cabbage and other Brassicas

Brevicoryne brassicae (L.)*, the cabbage aphid – A grey-green aphid with short siphunculi. The body is covered in a grey waxy powder. This aphid forms dense colonies often very wet with honeydew.

Myzus ascalonicus (Doncaster), the shallot aphid – A greyish-yellow aphid without a powdery covering and with short, swollen siphunculi (G.5).

G.5

Hogweed (*Heracleum sphondylium*)

Caveriella species*, the willow aphid – Medium-sized, greenish aphids, forming large colonies beneath the flowers. They migrate from willow to various umbelliferous plants in the summer. This genus is distinguished from other aphids by the possession of a supracaudal process which is a projection of the last abdominal segment above the cauda (G.6) and by having the siphunculi longer than the cauda.

G.6

C. *aegopodii* (Scopoli) – The supracaudal process is less than twice as long as the thicker, basal part of the last antennal segment.

C. *pastinacae* (L.) – The supracaudal process is about three times as long as the thicker basal part of the last antennal segment.

Aphis fabae Scopoli, the black bean aphid – Black or blackish-green aphids with pale legs and white wax markings on the body.

Hyadaphis foeniculi (Passerini), the honeysuckle aphid – A dark bluish-green aphid which can also be yellowish or brown, with long black siphunculi. This species is found on honeysuckle in the spring and migrates to umbellifers in the summer.

Lime trees (*Tilia* species)

Eucallipterus tiliae (L.), the lime aphid – A yellow-green aphid with distinctive black markings on the wings (G.7).

G.7

Oak (*Quercus* species)

Thelaxes dryophila (Shrank) – A reddish-brown to green aphid which is flattened and broadly oval in outline. Large individuals are thinly covered with white powder.

Tuberculoides annulatus (Hartig) – A small, yellow-green aphid. Each segment of the antennae is dark at the tip. The body is covered with powdery wax.

Myzocallis castinicola Baker – A yellow-green aphid. The body is without a wax covering.

Lachnus species – Large, brown, active aphids with short siphunculi.

Nettle (*Urtica dioica*)

Microlophium carnosum (Buckton)*, the nettle aphid – A large, green or pink, shiny aphid with long, narrow siphunculi and well-developed antennal tubercles. On nettles all year.

Aphis urticata – A small aphid with small antennal tubercles and short siphunculi. From May to June they are dark green and form densely-clustered colonies on the growing tips of nettles. From July onwards, the summer generations are yellowish and are scattered on the undersides of leaves.

Pines and other conifers

Adelgids – Woolly aphids distinguished by the mass of white flocculence on the bodies. Many form galls on conifers. Several species (see Carter, 1971).

Eulachnus agilis (Kaltenbach), the spotted pine aphid – A green aphid with hairs arising from reddish-brown spots giving this aphid a speckled appearance.

Schizolachnus pineti (F.), the grey pine aphid – A small, shiny, hairy aphid covered with white powder.

Cinara species – Large, dull, brown or black hairy aphids which walk away slowly if disturbed.

Red campion (*Silene dioica*)

Brachycaudus species* – Shiny, dark-reddish to black, rounded aphids with a short, rounded cauda which is only slightly longer than its basal width.

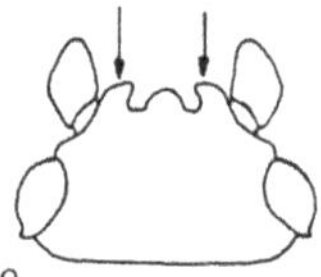

G.8

Myzus persicae (F.), the peach-potato aphid – A pale green to yellow aphid with convergent antennal tubercles (G.8).

Aphis fabae Scopoli, the black bean aphid – A black or blackish-green aphid, often with white markings on the abdomen. The cauda is longer than its basal width and is not shiny.

Reed (*Phragmites australis* [*communis*])

Hyalopterus pruni (Geoffroy)*, the mealy plum aphid – A green, narrow-bodied aphid, thinly covered with white powdery wax and with short siphunculi which are about as long as the cauda. This aphid forms small to large densely-aggregated colonies on the leaves from June to August.

Metopolophium dirhodum (Walker), the rose-grain aphid – A shiny green aphid without wax. It has long siphunculi and a dark green stripe along the middle of the back.

Roses (*Rosa* species)

Macrosiphum rosae (L.)*, the rose aphid – A large, pear-shaped green or pink aphid with long, black, tapering siphunculi, the colour of which distinguishes it from other rose aphids.

Macrosiphum euphorbiae (Thomas), the potato aphid – A long-bodied, greyish-green or pink aphid having long siphunculi which have a network pattern (G.9).

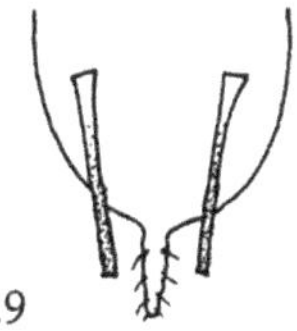

G.9

Metopolophium dirhodum (Walker), the rose-grain aphid – A shiny, green aphid with a dark green stripe along the middle of the back.

Myzaphis rosarum (Kaltenbach) – A green aphid with dark-tipped swollen siphunculi.

Pentatrichopus tetrarhodus (Walker) – A whitish-green aphid with numerous knobbed hairs arising from tubercles on the head and body.

Myzus ascalonicus (Doncaster), the shallot aphid – A greyish-yellow aphid with short, swollen, pale siphunculi.

Sycamore (*Acer pseudoplatanus*)

Drepanosiphum platanoidis (Shrank)*, the sycamore aphid – A large oval-shaped green aphid. It is found on sycamore all year round, but only breeds in the spring and late summer/autumn.

Periphyllus species – Blackish, hairy aphids with short siphunculi. Young stages are yellow-green and form tightly-packed groups.

Thistles (*Cirsium* species)

Brachycaudus species* – Shiny, dark-coloured, rounded aphids with caudas that are only slightly longer than their basal width.

Aphis species* – Dull black aphids with caudas that are longer than basal width.

Dactynotus species* – Large, reddish-brown to metallic-bronze aphids living on the stem, with a tendency to drop off if disturbed. *D. cirsii* has a yellow cauda and in *D. aeneus* the cauda is black.

Capitophorus species – Pale greenish or yellowish aphids with long siphunculi. The siphunculi are more than half as long as the hind tibia. *C. elaeagni* (Del Guercio) can be distinguished from *C. carduinus* (Walker) in that the former species has darkened tips to the siphunculi.

A guide to the common species of polyphagous aphids

Aphis fabae Scopoli, the black bean aphid – A dull-black or blackish-green aphid with pale markings on the legs. It sometimes has bar-like white wax markings on the abdomen. It is not easy to distinguish from other *Aphis* species except that *A. fabae* is reputedly the only black *Aphis* species to feed on beans (*Vicia faba*). The related *Aphis craccivora* (Koch) also feeds on beans but it is dark brown and black (see Stroyan, 1984 for further information on the genus *Aphis*).

G.10

Myzus persicae (F.), the peach potato aphid – A pale green to yellow aphid with convergent antennal tubercles and slightly swollen siphunculi (G.10). The winged individuals have a black patch on the abdomen.

G.11

Myzus ascalonicus (Doncaster), the shallot aphid – This is a greyish-yellow aphid with short, swollen siphunculi (G.11).

Myzus ornatus Laing, the violet aphid – A small, pale yellow or green aphid with paired dark markings on the thorax and abdomen.

Macrosiphum euphorbiae (Thomas), the potato aphid – A large, greyish-green or pink aphid with a network-like pattern on the long, narrow siphunculi.

Aulacorthum solani (Kaltenbach) – A medium-sized, greenish-yellow, shiny aphid with long siphunculi. At the base of each siphunculus there is a dark green patch.

< less than
> more than

Key I

Major groups of insects found at aphid colonies

1 Wings fully developed (but sometimes hidden beneath wing cases); the insect can fly — 2

– Wings absent or with non-functional wing-buds — 25

2 Two pairs of wings — 3

– Only one pair of wings: Diptera, two-winged flies — 7

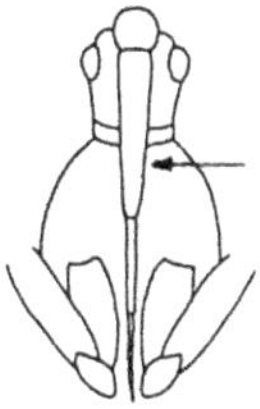

I.1

3 Mouthparts tubular, forming a needle-like, segmented probe (rostrum) (I.1, view from below) Hemiptera, bugs — 4

– Mouthparts of the biting type, with mandibles as I.2 — 19

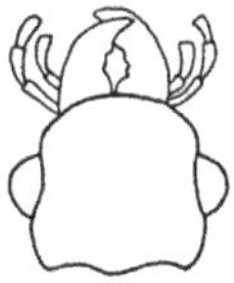

I.2

4 Rostrum of 3 segments, not pressed firmly against the body when the insect is resting (I.3, view from side or below); forewings with costal fracture and cuneus (I.4); small, <5mm long, red to dark-brown bugs with ocelli on the head (I.4) Anthocoridae, flower bugs (see Key V)

– Not like this — 5

I.3

5 Rostrum curved and not pressed against the body when at rest (I.5); wings without costal fracture and cuneus (I.6); >5mm long — Nabidae, Damsel bugs (see Southwood & Leston, 1959)

– Not like this — 6

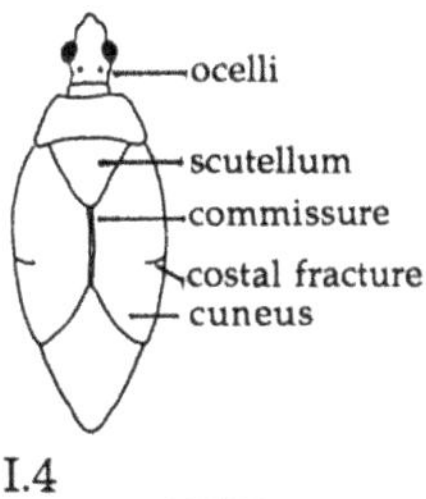

I.4

6 Rostrum straight, pressed against the body when at rest; forewings with costal fracture and cuneus; scutellum shorter than commissure (I.6) — Miridae, Capsid bugs (see Southwood & Leston, 1959)

– Not like this — other Hemiptera: Heteroptera (see Southwood & Leston,1959; Chinery, 1976)

I.5

7 Antennae with 5 or more segments — Nematocera, Thread horns — 8

– Antennae with <5 segments (usually 3), excluding a 2 – 3 jointed bristle (arista) on the third segment — 9

8 Antennae beaded, with circlets of hairs (I.7); wings usually with a fringe of hairs and few veins; first tarsal segment < 1/4 length of the second tarsal segment; very small (<4mm long) delicate flies — Cecidomyiidae, Gall midges including Aphid midges (see Key IV)

– Not like this — other Nematocera (see Unwin, 1981)

I.6

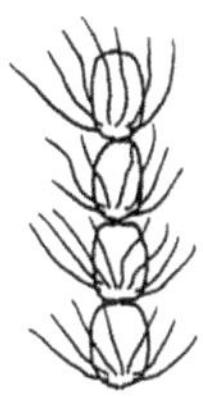

I.7

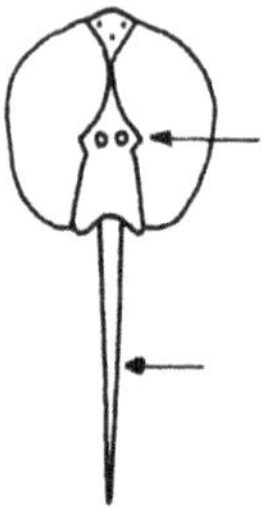

I.8

9 Head small and rounded with notches in the eyes at the front (I.8); small to large flies with needle-like mouthparts — Empididae, Dance flies

(Predatory flies; small species may feed on aphids and large species may take visitors to aphid colonies, see table 4)

– Not like this — 10

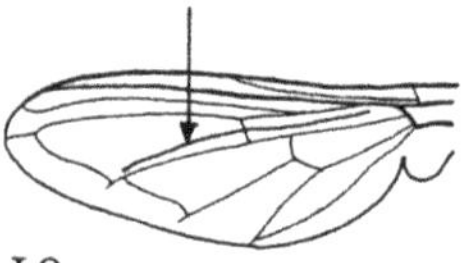

I.9

10 Vena spuria (not a true vein but a thickening in the wing, which passes through the anterior cross vein, I.9) present, abdomen often with yellow spots or bars like bees and wasps — Syrphidae, Hoverflies (see Key II)

– Vena spuria absent — 11

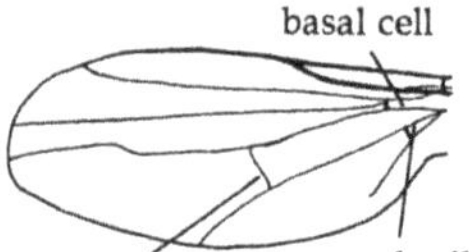

I.10

11 First basal and anal cells in the wings very short or absent (I.10); discal cell closed by a cross vein (I.10); tarsi with 2 pulvilli (I.11); metallic bluish, or green or yellow flies with long bristly legs; often with long heads in profile (I.12); small to large flies — Dolichopodidae

(Predatory flies, often seen at aphid colonies but prey ranges are largely unknown, see table 4)

– Not like this — 12

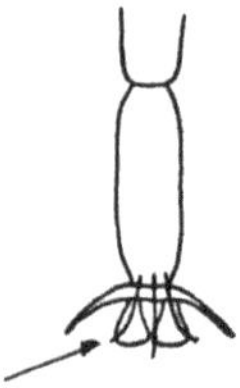

I.11

12 A large triangular mark on top of the head (I.13); small (<5mm long) black or yellow and black, shiny flies with few hairs — Chloropidae, Shoot flies — 13

– Not like this — 16

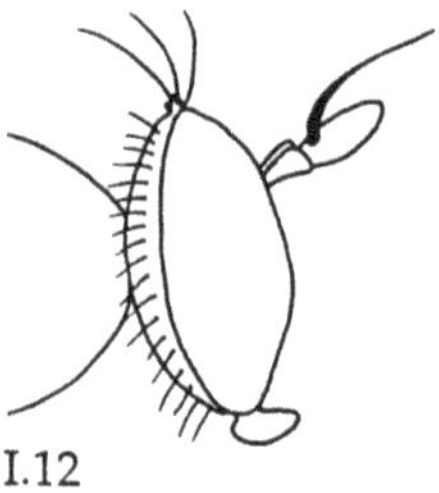

I.12

13 Upper surface of thorax with yellow and black stripes; hind femora normal not enlarged; scutellum (the semi-circular projection of the thorax that comes at the beginning of the abdomen, see II.36, p.52) flattened on upper surface — *Thaumatomyia* species — 14

(The larvae are predators of root aphids, see table 3; adult flies are often seen hovering close to aphid colonies; they feed on honeydew; the flies overwinter as adults and sometimes come into houses)

– Not like this — other Chloropidae

(There is no up to date guide to these flies which, except for *Thaumatomyia* species, are casual visitors to aphid colonies feeding on honeydew)

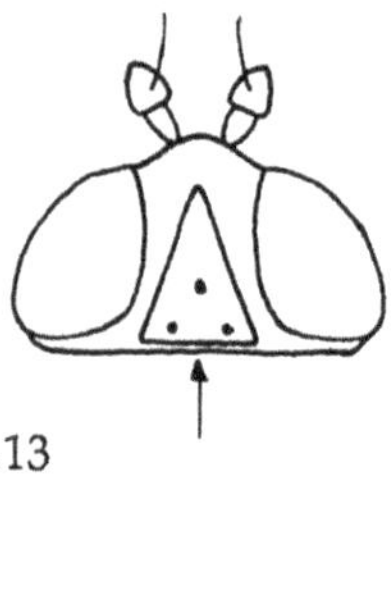

I.13

14 Flat part of the scutellum covered with bristles — 15

– Flat part of the scutellum bare except for a pair of long bristles at the tip — *Thaumatomyia* (*Chloropisca*) *glabra* (Meigen)

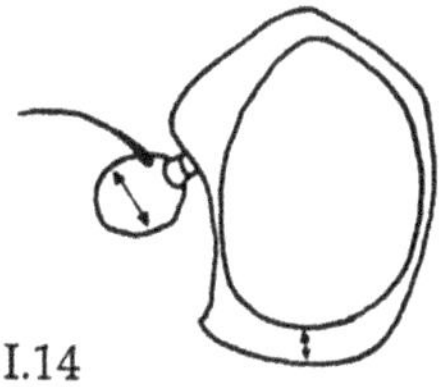

I.14

15 3rd antennal segment mostly black; jowls $^{1}/_{2}$ as wide as third antennal segment is deep (I.14)
Thaumatomyia notata (Meigen)

– Not like this other, rarer *Thaumatomyia* species

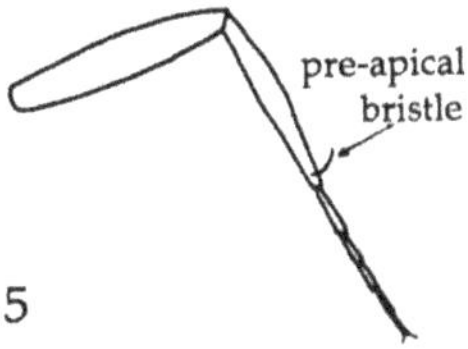

I. 15

16 Flies with long, thin legs; tibiae without pre-apical bristle (I.15);vibrissae (a pair of long, forwardly-directed bristles between the mouthparts and the eyes) absent; face receding (I.16); small to medium (up to 10mm long) flies Micropezidae, Stilt-legged flies (see table 4)

– Not like this 17

17 Small (<5mm long), heavily dusted, greyish flies with post-vertical bristles convergent (I.17) or these bristles absent; hind tibiae without a pre-apical bristle above
Chamaemyiidae (see table 3)

– Not like this 18

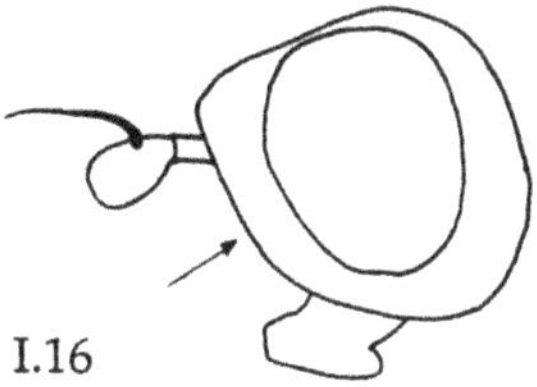

I.16

18 Medium-sized, hairy (5 – 10mm long) flies with round heads and pale hairs beneath (I.18)
Scathophagidae, Dung flies

(Some species in this family are predatory, possibly on aphids but more probably on larger insects. For instance, the common yellow dung fly, *Scathophaga stercoraria* (L.) is often present at aphid colonies attempting to catch insect visitors for prey.)

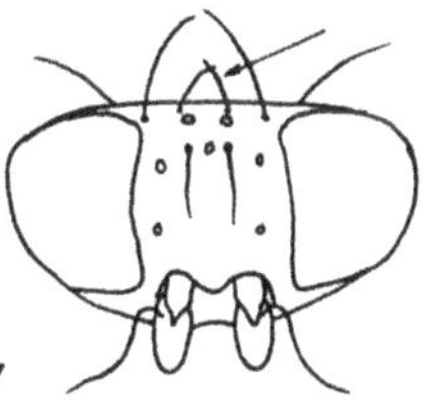

I.17

– Not like this other Diptera

(See Colyer & Hammond, 1968 and Unwin, 1981. Large numbers of flies visit aphid colonies to feed on honeydew. Particularly noticeable at large aphid colonies are bluebottles and their allies [Calliphoridae])

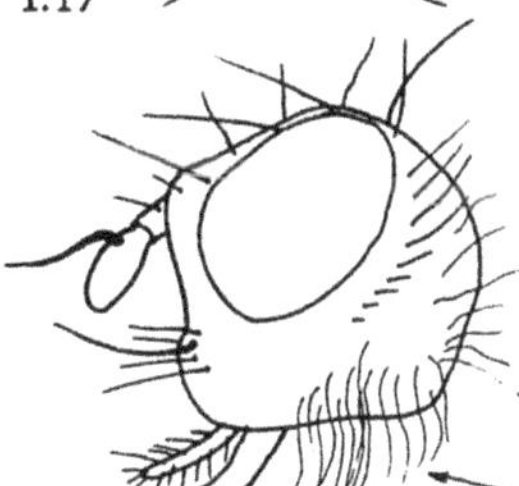

I.18

Insects with wings and biting mouthparts

19 Body ending in curved forceps (I.19); mostly night-time visitors to aphid colonies Dermaptera, Earwigs (see p. 15)

– Not like this 20

20 Hard wing cases hide the functional wings
Coleoptera, Beetles 21

– No hard wing cases 24

21 Antennae with enlarged tips like a club; antennae as long as, or shorter than, the width of the thorax; tarsi with 3 segments, some of which are two-lobed (I.20); mostly oval beetles less than 10mm long with red, yellow or black spots
Coccinellidae, Ladybirds (see Key III)

– Not like this 22

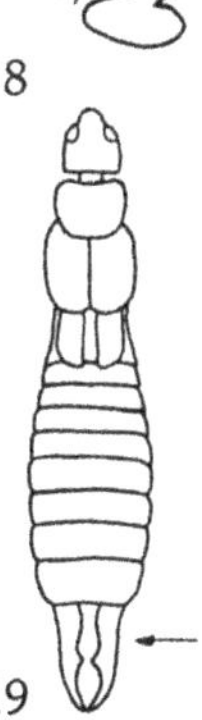

I.19

I.20

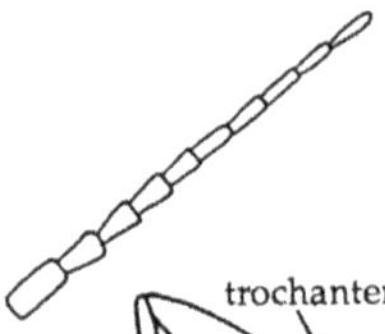

I.21

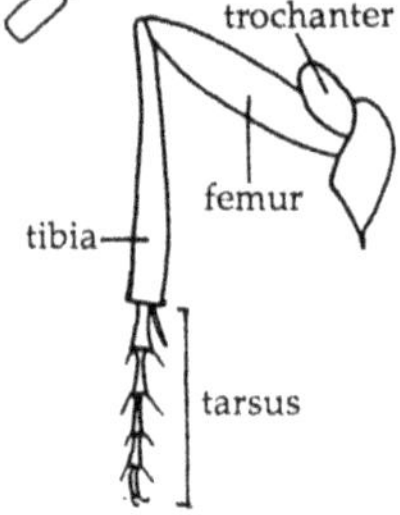

I.22

22 Antennae threadlike (I.21); wing cases covering all or most of the abdomen; tarsi with 5 segments; hind trochanters strongly projecting and longer than the diameter of the hind femur (I.22); small to large beetles Carabidae, Ground beetles (see p. 63)
– Not like this 23

23 Wing cases shorter than the exposed part of the abdomen (I.23); hind margins of the wing cases meet on the mid-line (I.23); tarsi with 5 or, less frequently, 4 segments Staphylinidae, Rove beetles (see p. 63)
– Not like this other beetles (see Unwin, 1984)

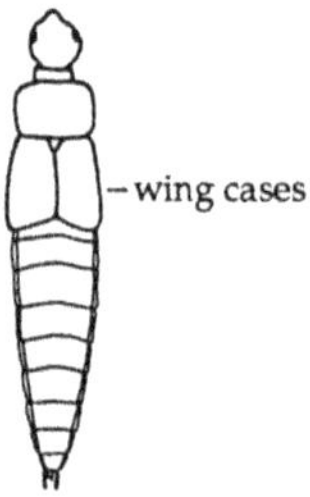

I.23

24 Large transparent wings often longer than the abdomen with a complex network of veins (I.24); wings held roof-like over the abdomen when the insect is resting; abdomen long and narrow without a waist (narrow portion between the thorax and abdomen); antennae long and simple (composed of similar segments one after the other); variously coloured insects but not shiny black Neuroptera, Lacewings (see Key VI)
– Wings usually shorter than the abdomen, without such a complex network of veins; forewings 'zip' on to the hind wings in flight
Hymenoptera (bees, wasps and parasitoids)
(See Yeo & Corbet (1983) for details of aphid-hunting wasps, e.g. *Passaloecus*; for further details of Hymenoptera see Chinery (1976).)

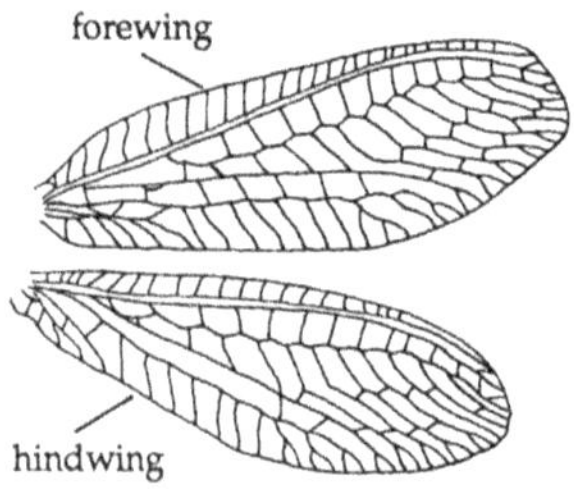

I.24

Insects with non-functional wing buds or wings absent

(Some common rove beetles e.g. *Pterostichus* have very short wings and are unable to fly – refer to couplet 2 in such cases.)

25 Red, brown or black, fast-moving insects with a narrow waist between the abdomen and thorax (I.25); usually many individuals present at the same time
Formicidae, Ants (see Key VII)
– Not like this 26

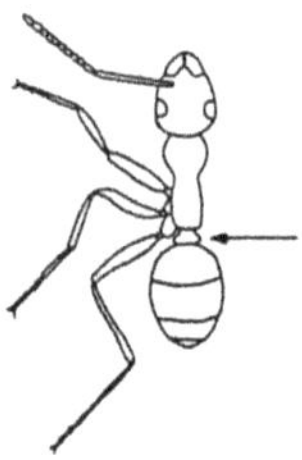

I.25

26 Larva with 3 pairs of legs 27
– Larva without legs 35

27 Mouthparts tubular, forming a needle-like segmented probe (the rostrum) (I.26) (view from side or below) 28
– Rostrum absent 30

I.26

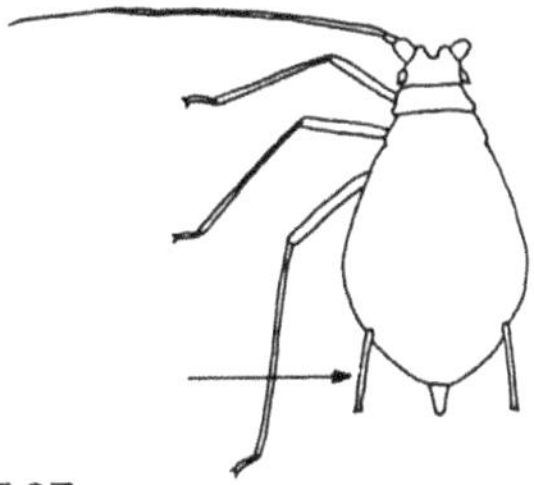

I.27

28 Siphunculi present on the tip of the abdomen (I.27)
Aphids (see guide p. 37)

– Siphunculi not present 29

29 Rostrum consists of 3 segments and is not pressed firmly against the underside of the body when the insect is resting (view from the side); shining red to dark brown insects; <5mm long Anthocoridae (Flower bugs)
(see pl. 1.6 and Key V)

– Not like this other Hemiptera: Heteroptera, Bugs
(see Southwood & Leston, 1959)

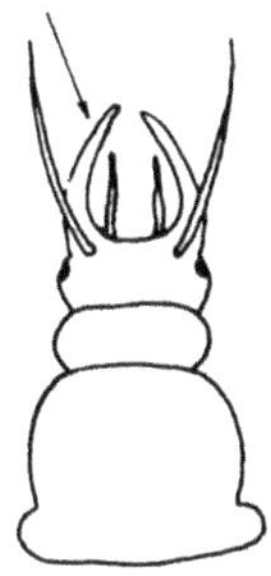

I.28

30 Head with long, curved mandibles (I.28)
Neuroptera (Lacewings)
(see pl. 1.2 and Key VI)

– Head without long, curved mandibles 31

31 Body slate-grey to black or yellowish with row of shining black spots and occasional pale markings; conspicuous, thick, black legs Coccinellidae, Ladybirds
(see pl. 1.3 and Key III)

– Not like this 32

32 Abdomen with prolegs; body variously coloured and patterned Caterpillars, Lepidoptera, butterflies and moths, and Symphyta, sawflies
(Lepidoptera have five pairs of prolegs including the terminal pair. Sawflies have six or more.)

– Abdomen without prolegs Coleoptera (beetle) larvae 33

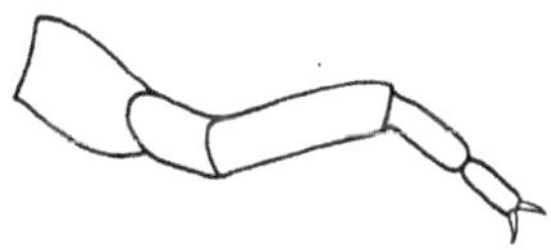

I.29

33 Body whitish, flattened or ovoid and straight when the insect is resting; active larvae with biting mouthparts and a darkened head capsule; three pairs of thoracic legs present 34

– Not like this other Coleoptera (beetle) larvae
(see van Emden, 1942)

34 Thoracic legs with five segments (I.29)
Carabidae (Ground beetles), (see p. 63)

– Thoracic legs with four segments (I.30)
Staphylinidae (Rove beetles), (see p. 63)

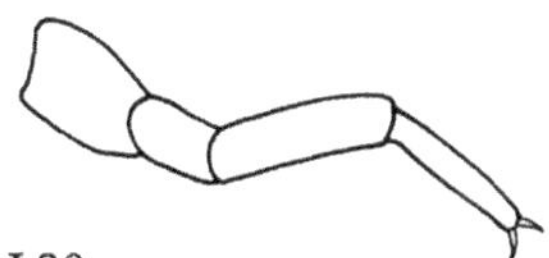

I.30

35 Small (up to 4mm long), slow-moving, shiny, pink to orange-red larvae without a pair of obvious tawny-brown or black ovoid plates at the end of the body
Cecidomyiidae, Gall midges including Aphid midges
(see pl. 1.4 and Key IV)

– Not like this 36

I.31

36 Larvae with a conspicuous tawny-brown or black breathing tube at the end of the body (I.31); active larvae often patterned and brightly coloured; segmentation not obvious owing to a smooth surface or wrinkles and folds between the segments — Syrphidae, Hoverflies (see pl. 1.1 and Key II)

– At the end of the body the breathing tube is separated into two parts; segmentation obvious; <6mm long — 37

37 Grey dusted larvae with many small projections; up to 6mm long; dark collar or ring behind the head — Chamaemyiidae (see Table 3)

– Not like this — other Diptera (True flies) larvae (See Table 3 for the root-dwelling aphid predators, *Thaumatomyia* (Chloropidae) and Phoridae. Small (<6 mm) very translucent larvae with breathing tube in two parts are probably 1st or 2nd stage hoverfly larvae)

Key II
Hoverflies, Syrphidae
Larvae

II.1

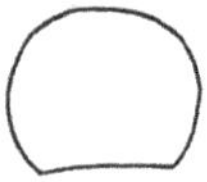
II.2

II.3

II.4

II.5

1 Larva green or pink — 2

– Larva otherwise coloured — 7

2 In the rear half of the body, larva flattened in cross section (II.1) — 3

– In the rear half of the body, larva rounded (II.2), oval (II.3) or rectangular with side ridges (II.4) in cross section — 4

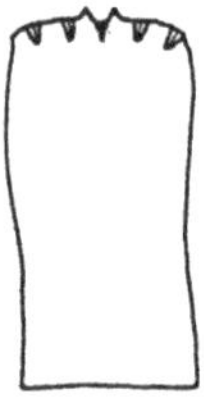
II.6

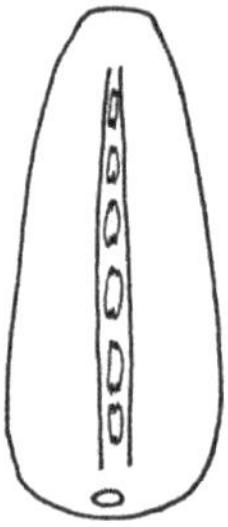
II.7

II.8

3 White or cream stripe down the middle of the upper surface without a chain link pattern (II.5); posterior respiratory process twice as long as its width at the tip (II.6), May to July — *Epistrophe eligans* (Harris)

– Stripe down the middle of the upper surface with a chain link pattern (II.7); posterior respiratory process about as broad at the tip as long (II.8); August to October — *Epistrophe grossulariae* (Meigen)

II.9

4 Rectangular in cross section (II.4); mottled pink, brown and white with arrow-head markings on the upper surface (II.9); a pair of ridges on the sides of the upper surface of the body (view from above at low magnification and tilt the larva from side to side in the light); May to October — *Platycheirus scutatus* (Meigen)

– Rounded or oval in cross section and without ridges running along the sides at the top of the body — 5

II.10

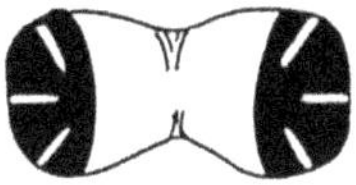
II.11

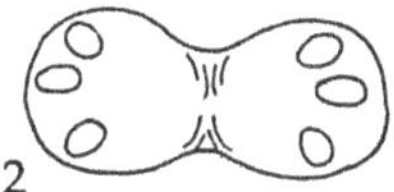
II.12

II.13

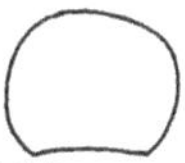
II.14

II.15

II.16

II.17

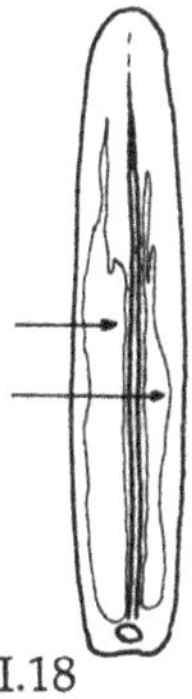
II.18

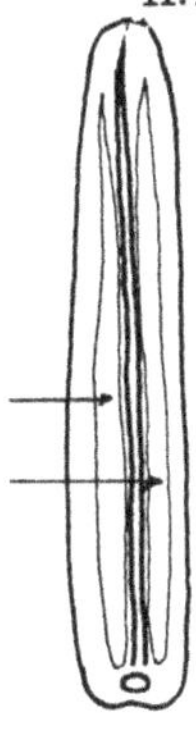
II.19

5 Posterior respiratory process set in a triangular hollow with the spiracles set on ridges (II.10); green or pink with a white or cream stripe along the upper surface; body rounded in cross section; large, up to 16mm long, July to September *Scaeva pyrastri* (L.)

– Posterior respiratory process not set in a triangular hollow; body oval in cross section; up to 12mm long 6

6 Posterior respiratory process one and a half times as long as its width at the base with a pair of black ovoid markings surrounding the spiracles (II.11); bright green; late July to October *Spaerophoria scripta* (L.)

Note: In the spring (April to June) the less common *Spaerophoria menthastri* (L.) may be recorded. It can be distinguished from *S. scripta* in that the former species has the posterior respiratory process as long as its width at the base.

– Posterior respiratory process twice as broad as long and not black-tipped (II.12); shining translucent, green larvae; May to November *Melanostoma scalare* (F.)

7 In the rear half of the body, larva rectangular in cross section (II.13); a pair of ridges present on the sides of the upper surface of the body (view from above at about x 10 and tilt the larva from side to side in the light) 8

– Body round (II.14), oval (II.15) or flat (II.16) in cross section; without ridges along the sides of the upper surface 11

8 A series of 4 or 5 arrow-head marks on the upper surface (II.17) 9

– Arrow-head marks absent 10

9 Body partially translucent so that the internal organs are visible between the light covering of brown and white fat bodies other *Platycheirus* species

– Body not translucent, because many brown, pink, green and white particles of fat completely obscure the interior contents of the larva; October to May overwintering form of *Platycheirus scutatus*

10 Larva with a non-symmetrical pattern of fat bodies in the two halves of the upper surface (II.18); body pale yellowish-brown and narrow *Baccha* species

– Larva with a symmetrical pattern of fat bodies in the two halves of the upper surface (II.19); body pale or dark other *Platycheirus* species

11 Larva flat in cross section 12
– Larva rounded or oval in cross section 17

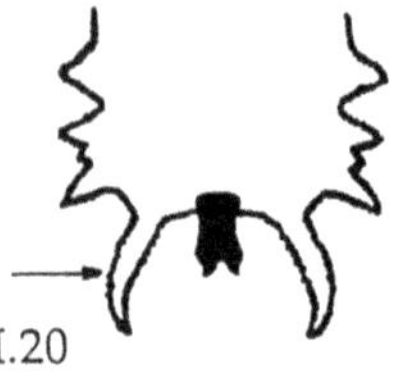
II.20

12 A pair of long, backwardly-directed spines on the last segment of the body (II.20); body mottled dark grey and black 13
– Long spines absent on the last segment of the body 14

II.21

13 Tips of the posterior respiratory process separated by more than the length of a spiracular plate (II.21); May to October *Dasysyrphus albostriatus* (Fallén)
– Tips of the posterior respiratory process separated by less than the length of a spiracular plate (II.22); May to October *Dasysyrphus tricinctus* (Fallén)

14 Larva all white or white and broadly-edged red-brown; not translucent; July to August *Melangyna umbellatarum* (F.)
– Larva partially translucent; with stripes, arrow-head shaped markings or other markings 15

II.22

15 Pale orange-brown larva with a white stripe down the middle of the upper surface; markedly flattened larva overwintering form of species of *Epistrophe* larvae (see couplet 3, p. 48)
– Larva otherwise coloured, without a white stripe down the middle of the upper surface 16

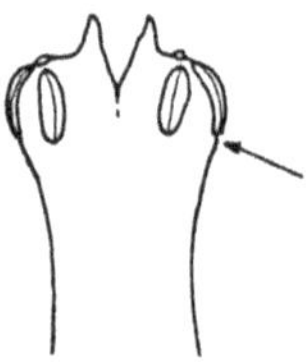
II.23

16 Translucent larva with white arrow-head shaped markings on the upper surface; posterior respiratory process with a rounded tip (view from the front) (II.23); common at aphid colonies on beech trees, more rarely on other deciduous trees; May to July *Meligramma cincta* (Fallén)
– Larva otherwise coloured and marked other syrphid species

(Rear and identify the adult fly; see below for key to adult hoverflies and Stubbs & Falk, 1983)

II.24

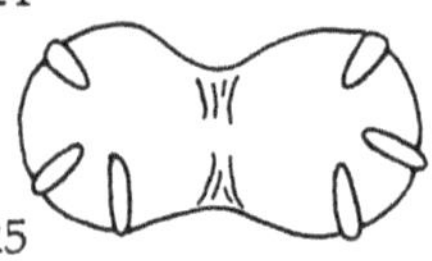
II.25

17 Spiracles not extending over the sides of the posterior respiratory process when viewed from above (II.24); a broad, ovoid, black or white mark round the spiracles (II.24) *Platycheirus* or *Baccha* species, follow couplet 8
– Spiracles extending over the sides of the posterior respiratory process (II.25); no ovoid, black or white mark around the spiracles (II.25) 18

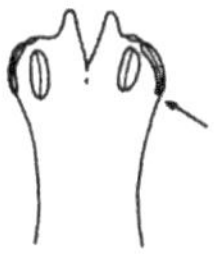

II.26

II.27

18 Posterior respiratory process with a rounded tip (view from the front) (II.26) *Meligramma cincta*, see couplet 16
– Posterior respiratory process without a rounded tip 19

19 Posterior respiratory process set in a triangular hollow (view from behind) (II.27); body mottled brown and white and surface patchily covered in little black spikes (to view use at least x 20) *Metasyrphus* species 20
– Posterior respiratory process not set in a triangular hollow; body otherwise coloured and without little spikes 21

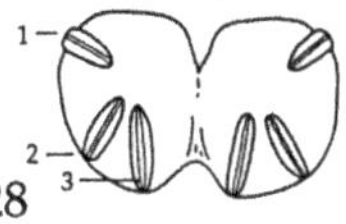

II.28

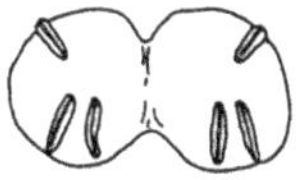

II.29

20 Spiracles 2 and 3 diverging (II.28); May to October *Metasyrphus luniger* (Meigen)
– Spiracles 2 and 3 almost parallel (II.29); May to August *Metasyrphus corollae* (F.)

Note: Separating these species can be difficult; if in doubt rear and identify the adult.

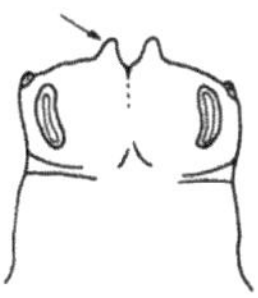

II.30

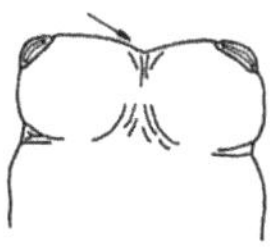

II.31

21 The tip of the posterior respiratory process with dorsal spurs (II.30) *Syrphus* 22
– Dorsal spurs absent (II.31); shining translucent larva with an asymmetrical pattern of white fat bodies in the two halves of the upper surface; bright red tubules visible in the rear half of the body (view from above); May to August, most common in late July and August *Episyrphus balteatus* (Degeer)

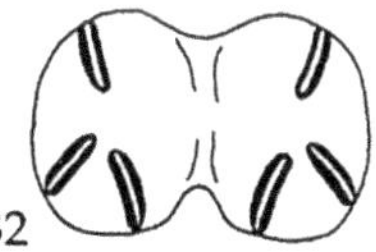

II.32

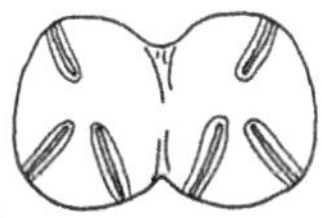

II.33

22 Spiracles broad and diffusely black-bordered (use at least x 20 to view) (II.32); May to November *Syrphus torvus* Osten-Sacken
– Spiracles brown-bordered or narrow and sharply black-bordered (II.33) 23

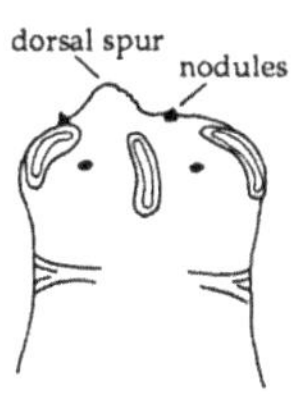

II.34

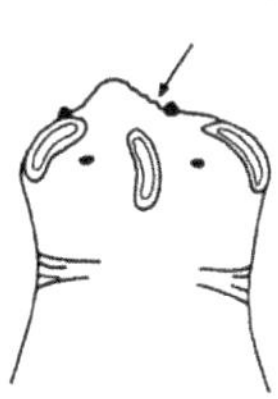

II.35

23 Posterior respiratory process pale brown with a width at the tip of less than 0.55mm; dorsal spur not reaching nodule (view from the side) (II.34); May to November *Syrphus vitripennis* Meigen
– Posterior respiratory process tawny-brown with a width at the tip of more than 0.55mm; dorsal spur almost reaching nodule (view from the side) (II.35); May to November *Syrphus ribesii* (L.) (pl. 1.1)

Note: *Syrphus* species can be difficult to separate. Although the differences outlined above should separate the majority of *Syrphus* larvae, intermediates may occur. If in doubt, rear and identify the adult. *Syrphus ribesii* is often the commonest hoverfly larva at many aphid colonies.

Adult hoverflies

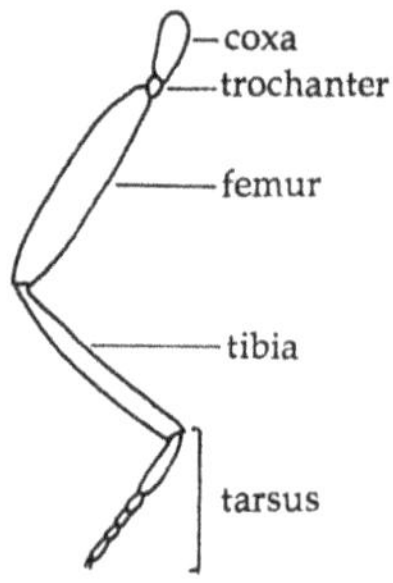

a hoverfly leg

This key should be used only to identify reared specimens. There are nearly 100 species of aphidophagous hoverflies and many of the adults are more commonly recorded than larvae. The species included in the key have been selected because they comprise over 95% of the larvae usually recorded at the aphid colonies given in the guide. For descriptions, keys and pictures of other hoverfly species, Gilbert (1986) and Stubbs & Falk (1983) should be consulted. Male aphidophagous hoverflies differ from females in that they have eyes which meet at the top of the head. In females, the eyes are separated.

1 Face (region between the eyes and below the antennae) black 2
– Face yellow or yellow with central black line or blotch 5

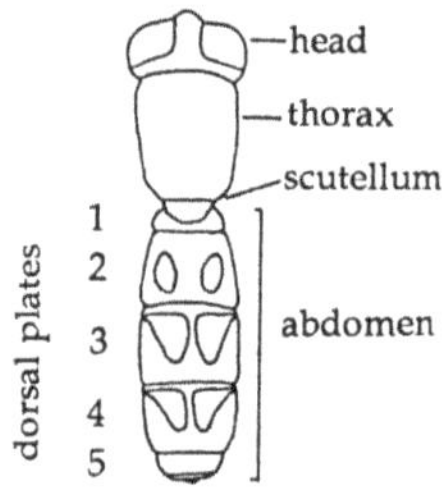

II.36

2 Abdomen very thin and needle like (at narrowest, no wider than scutellum) *Baccha* species
– Abdomen wider 3

3 Male front tibiae and tarsi without enlarged or flattened sections; female with triangular, orange markings on dorsal plates 3 and 4 (II. 36) *Melanostoma scalare*
– Male front tibiae and tarsi with enlarged and flattened sections (II.37); female with oval orange or silver markings on dorsal plates 3 and 4 4

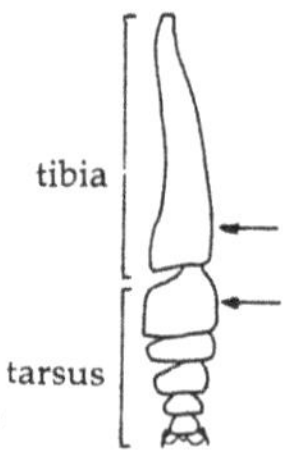

II.37

4 Male front tibiae with a tuft of black hairs on the outer edge (II.38); female with all the following characters:
(i) yellowish mark on the underside of the last segment of the antennae,
(ii) dorsal plates with yellow, not silver, markings,
(iii) hind margins of yellow marks on dorsal plate 2 straight not oblique *Platycheirus scutatus*
– Male front tibiae without a tuft of black hairs; female not with all the characters, i – iii, together other *Platycheirus* species

II.38

5 Long hairs on the upper surface of the squamae (pale flaps at the base of the wing) (II.39) *Syrphus* 6
– No hairs on the upper surface of the squamae 8

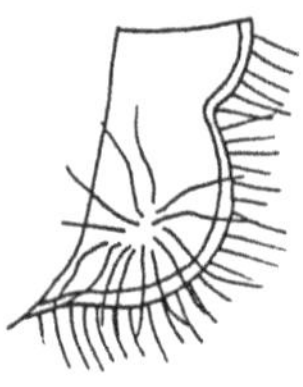

II.39

6 Eyes coated with short hairs (view from above alternately against a pale and dark background using at least x 10) *Syrphus torvus*
– Eyes bare, or with a few hairs present 7

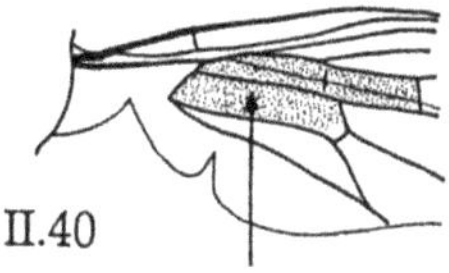

II.40

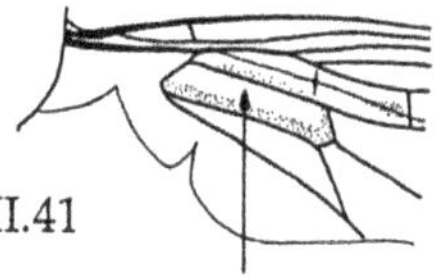

II.41

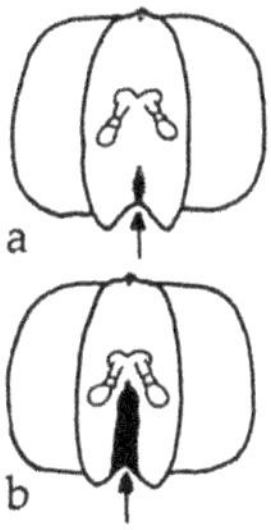

a

II.42 b

7 Male basal cells of wing covered in short, black hairs (II.40); female hind femora yellow except at extreme base *Syrphus ribesii*

– Male basal cells of wing with areas bare of black hairs (II.41); female hind femora black with a yellow tip *Syrphus vitripennis*

8 Eyes covered in short hairs (view from above alternately against a pale and dark background using at least x 10) 9

– Eyes bare 12

9 Hind tibiae mostly black *Melangyna umbellatarum*

– Hind tibiae mostly yellow or yellow with a black central band 10

10 Face pale yellow with a short and narrow dark central stripe (II.42a); abdomen with whitish, comma-shaped markings *Scaeva pyrastri*

– Face dark yellow with a long and broad central stripe (II.42b); abdomen with yellow stripes or bars 11

11 Markings on dorsal plate 3 wider than those on other dorsal plates *Dasysyrphus tricinctus*

– Markings on dorsal plate 3 not wider than those elsewhere *Dasysyrphus albostriatus*

12 Hind tibia mostly black *Melangyna umbellatarum*

– Hind tibia mostly yellow or with a black central band 13

13 Dorsal plate 2 with a pair of small, sharply-pointed triangular yellow marks and wide bands on dorsal plates 3 and 4 *Meligramma cincta*

– Second dorsal plate without small triangular markings 14

14 Very clear, shining yellow markings forming an uninterrupted stripe on each side of the top of the thorax *Sphaerophoria* 15

– Yellow markings absent or, if present, dull, not shining and separated 16

15 Male abdomen longer than the length of a wing *Sphaerophoria scripta*

– Male abdomen about as long as the length of a wing other *Sphaerophoria* species

16 Dorsal plates 3 and 4 with an upper and lower black band (the upper band is sometimes divided into a pair of narrow stripes) *Episyrphus balteatus*

– Dorsal plates 3 and 4 otherwise marked 17

17 Line of hairs on the side margins of the dorsal plates of the abdomen all black 18

– Pale hairs on the side margins of the dorsal plates 19

18 The yellow markings on dorsal plates 2 and 3 cover more than half of the side margins (view from above) *Metasyrphus corollae*

– Yellow markings on dorsal plates 2 and 3 cover less than half of the side margins *Metasyrphus luniger*

19 Dorsal plates 3 and 4 with yellow markings absent or narrower than those on dorsal plate 2 *Epistrophe eligans*

– Dorsal plates 3 and 4 with equally wide yellow bands *Epistrophe grossulariae*

Key III
Ladybirds, Coccinellidae
Larvae

In this book, the scientific names of ladybirds are written in conventional abbreviated form. Thus for instance *Coccinella septempunctata* appears as *Coccinella 7-punctata*. For further information on the identification of ladybirds, see Majerus & Kearns (1989), and Hodek (1973).

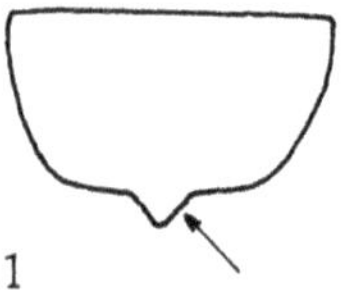

III.1

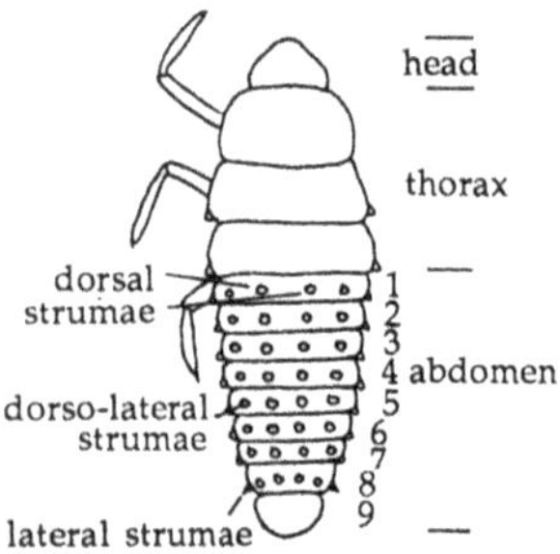

III.2

1 9th (last) abdominal segment with a blunt projection (III.1); (white markings on thorax and abdomen) 14 spot ladybird, *Propylea 14-punctata* (L.)

– Last abdominal segment without a blunt projection 2

2 Dorso-lateral strumae on dorsal plate 4 dark (III.2) 3

– Dorso-lateral strumae on dorsal plate 4 pale 4

3 Strumae at sides of dorsal plates 2, 3 and 5 – 8 dark; ground colour ('skin' colour, beneath the body markings) grey to black
2 spot ladybird, *Adalia 2-punctata* (L.)
(*Adalia 2-punctata* colour varieties are not uncommon, so if in doubt rear and identify the adult.)

– Strumae at sides of dorsal plates 3 – 8 whitish; ground colour light grey to grey
10 spot ladybird, *Adalia 10-punctata* (L.)

III.3

4 Base of claw at the tip of the legs rounded without a squarish tooth (III.3); centre of head pale, up to 6mm long 11 spot ladybird, *Coccinella 11-punctata* L.

– Base of claw at the tip of the legs with a well-developed squarish tooth (III.4) 5

III.4

5 Dorso-lateral strumae on dorsal plates 6 and 7 orange-red; up to 6mm long
5 spot ladybird, *Coccinella 5-punctata* L.

– Dorso-lateral strumae on dorsal plates 6 and 7 black; usually >6mm long
7 spot ladybird, *Coccinella 7-punctata* L.

See Majerus & Kearns (1989) or Hodek (1973) for the identification of other ladybird larvae.

Adult ladybirds

1 Wing cases (elytra) red with black spots, all black, or black with red spots 2

– Wing cases yellow with black markings, or rarely, mostly black with yellow markings 6

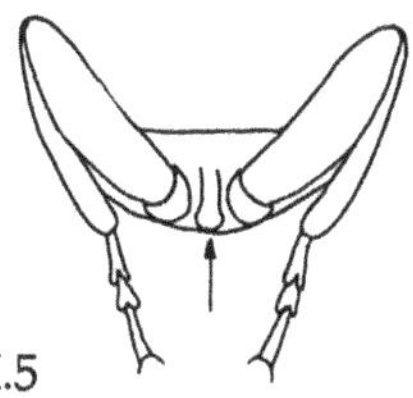

III.5

2 Ridges between the front legs (III.5) (view from beneath using at least x 20; if hairs are in the way use a pin to move them aside) *Coccinella* 3

– Ridges absent between the front legs *Adalia* 5

3 Wing cases with fewer than 11 spots in total including the scutellary spot (spot at the base of the wing cases where they meet in the mid-line); ridges between the front legs parallel 4

– Wing cases with a total of 11 spots including the scutellary spot; ridges between front legs converge in front *Coccinella 11-punctata*

4 Wing cases with a total of 5 spots including the scutellary spot; small species (3.5 – 4.5mm long) *Coccinella 5-punctata*

– Wing cases with 7 spots including the scutellary spot; large species (5.5 – 7.5mm long) *Coccinella 7-punctata*

5 A yellow marking above and to the side of the second pair of legs (view from beneath); elytra red with 5 small black spots on each (but many varieties occur with all-red to all-black elytra); 3 – 4mm long *Adalia 10-punctata*

– No yellow marking above and to the side of the second pair of legs (view from beneath); elytra red with 1 black spot on each (but many varieties occur from all-red to all-black elytra and various combinations of spots and markings) *Adalia 2-punctata*

(See Linssen, 1959 or Majerus & Kearns, 1989 for details of colour pattern varieties in *Adalia* ladybirds.)

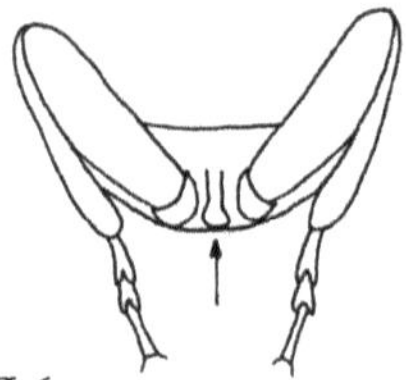
III.6

6 A pair of ridges between the front legs (III.6) (view from beneath using at least x 20; if hairs are in the way use a pin to move them aside) 7

– Ridges absent between the front legs; wing cases yellow with rectangular black markings, some of these fused together *Propylea 14-punctata*

7 Wing cases yellow with at least 11 black spots on each; small species (2 – 3mm long); scutellary spot absent *Psyllobora 22-punctata* (L.)

(Both adults and larvae of this species feed on mildews. The adults are, however, sometimes seen close to aphid colonies. Whether these adults are feeding on aphids or honeydew is not known for certain.)

– Wing cases with fewer than 11 spots on each; scutellary spot present; larger species (3 – 4mm long) *Adalia* colour variety, see couplet 5

See Majerus & Kearns (1989) or Pope (1953) for identification of other ladybird species.

Key IV
Aphid midges, Cecidomyiidae
Larvae

1 Sternal spatula present (IV.1) (turn larva over to view and use at least x 20); segmentally arranged ridges present on the underside of the body (IV.1) 2

– Not as above other Cecidomyiidae

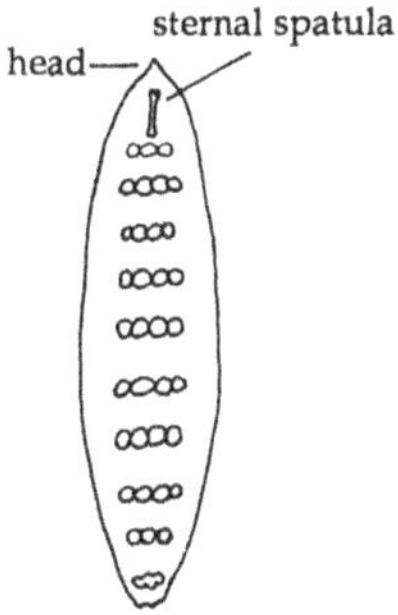

IV.1

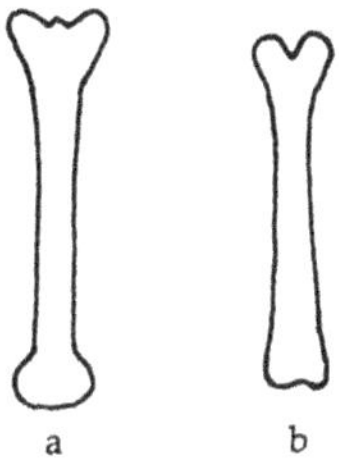

IV.2

e.g. *Lestodiplosis* larvae which are chiefly predators of other cecid larvae, and do not possess a sternal spatula, or *Mycodiplosis* larvae which feed on fungi associated with honeydew and are distinguished by having a broad head which is hinged downwards.

2 Abdomen with numerous hairs, obvious even at low magnification *Monobremia subterranea* (Kieffer)
– Abdomen with only inconspicuous hairs 3

3 Sternal spatula wider and enlarged at the rear end (IV.2a) *Aphidoletes aphidomyza* (Rondani)
– Sternal spatula narrow, without an enlarged rear end (IV.2b) *Aphidoletes urticariae* (Kieffer)

Adult aphid midges

Adult aphid midges are not easy to distinguish. Harris (1966) uses male genital characters to separate *A. aphidomyza* from *A. urticariae.* The pattern of hairs on the outer section of vein 5 in the wings might separate the females. For further details see Harris (1966, 1973).

Key V
Flower bugs, Anthocoridae

< less than
> more than

Larvae

1 Legs, rostrum (beak) and antennae pale yellow *Orius* species (see Southwood & Leston, 1959)
– Legs, rostrum and antennae partly or completely dark reddish *Anthocoris* species 2

2 Wing buds pale at tips 3
– Wing buds not pale at tips 6

3 Rear angles of pronotum (see V.1) pale 4
– Rear angles of pronotum not pale 5

4 Antennae as long as greatest width of abdomen with antennal segments 2 and 3 pale, 1 and 4 dark; tibiae (p. 52) dark at base and at the tip *Anthocoris nemorum* (L.)
– Antennae shorter than width of abdomen, with antennal segments 2 – 4 pale at their tips; tibiae pale *Anthocoris nemoralis* (F.)

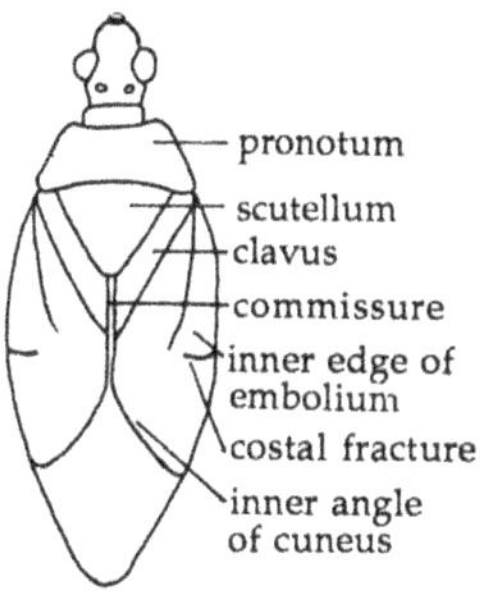

V.1

5 Length (tip of head between the antennae to the end of the abdomen) <3.5mm *Anthocoris confusus* Reuter
– Length >4mm other *Anthocoris* species (see Sands, 1957)

6 Length <3.5mm 7
– Length >4mm other *Anthocoris* species (see Sands, 1957)

7 Legs yellow-brown; antennae with pale markings *Anthocoris confusus* Reuter
– Legs dark brownish-red; antennae dark brown except at the extreme ends of each segment other *Anthocoris* species (see Sands, 1957)

Adult Flower bugs

1 Wings not fully formed, appearing as wing buds larval Anthocoridae, see above
– Wings fully formed 2

2 Small species (<3mm long); scutellum longer than commissure *Orius* species (see Southwood & Leston, 1959)
– Larger species; scutellum and commissure (V.1) approximately equal in length 3

3 Forewings entirely shining with brown tips and black areas, pronotum black *Anthocoris nemorum*
– Forewings partly dull especially the clavus and not patterned as above 4

4 Inner angle of cuneus dull; embolium with a dull band along its inner edge *Anthocoris confusus*
– Inner angle of cuneus and embolium shining *Anthocoris nemoralis*

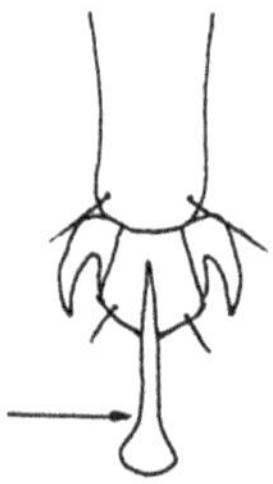

VI.1

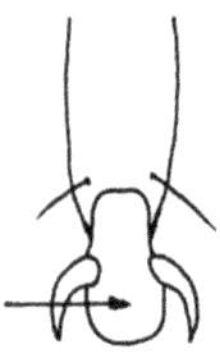

VI.2

Key VI
Lacewings, Neuroptera

Larvae

1 Empodium (grasping organ at the tip of the leg) trumpet-shaped (VI.1); prominent hairy bumps along the sides of the body; mandibles longer than the head, slender and strongly curved Chrysopidae 5
– Empodium short and rounded (VI.2); body without hairy bumps; mandibles short, thickened and less strongly curved Hemerobiidae 2

2 Triangular dark blotches on the body 3

– Triangular markings absent 4

3 Wedge-shaped marking on the surface of the head often extending to the front edge of the head, colouring light *Hemerobius lutescens* F.

– Wedge-shaped mark on head not usually reaching front margin, colouring darkish *Hemerobius humulinus* L.

Note: These characters are not always reliable because intermediates are sometimes found. Therefore larvae should be reared to the adult stage to confirm identity.

4 Head dark brown; body greyish; legs grey *Kimminsia subnebulosa* (Stephen)

– Head yellow with 3 black lines; legs yellow *Micromus variegatus* (F.)

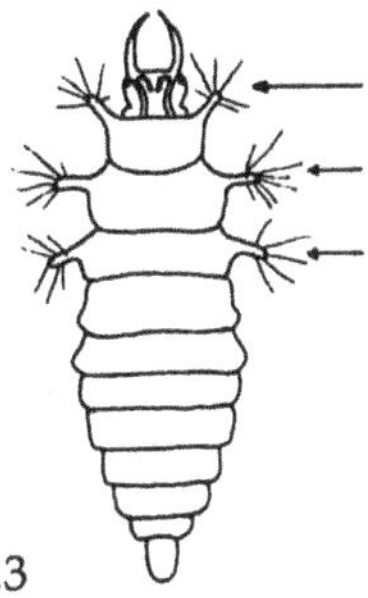

VI.3

5 Hairy bumps on the thorax at least 3 times as long as broad (VI.3); (body broad; larva covered with a layer of debris) *Chrysotropia ciliata* Wesmael

– Hairy bumps on the thorax short or absent 6

6 Body coated with a layer of debris *Anisochrysa ventralis* (Curtis)

– Body without debris 7

VI.4

7 Head pale with any dark lines or blotches pale and indistinct; body bright yellow with red markings on thorax and abdomen *Nineta vittata* (Wesmael)

– Head dark, or with distinct black lines and blotches 8

8 Hairy bumps and side margins of metathorax (VI.4) black or brown; no large black blotches on the head, although narrow black lines may be present 9

– Hairy bumps and side margins of metathorax pale; rear half of the head with large black blotches 10

9 Hairy bumps on the side of the body small and inconspicuous; no broad dark bands running along the body (pl. 1.2) *Nineta flava* (Scopoli)

– Hairy bumps on the side of the body prominent and conspicuous; body with 2 broad dark bands *Chrysoperla carnea* (Stephens)

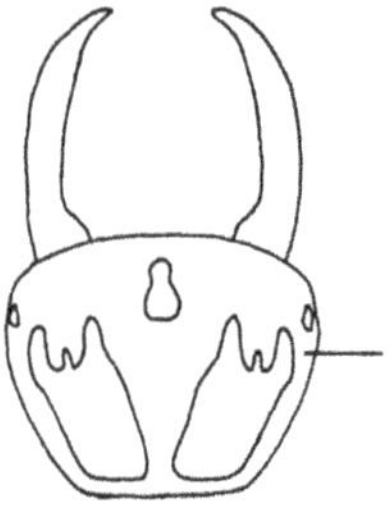

VI.5

10 Finger-like projections on the front edge of the black markings on the head (VI.5) *Chrysopa perla* (L.)

– No finger-like projections on the black markings of the head *Chrysopa septempunctata* Wesmael

Adult lacewings

Note: this key should be used to identify reared specimens only.

1 Small, <20mm long; brown or greyish species; antennae like a string of beads Hemerobiidae 2

– Larger, >20mm long; yellow or green species; antennae thread-like Chrysopidae 5

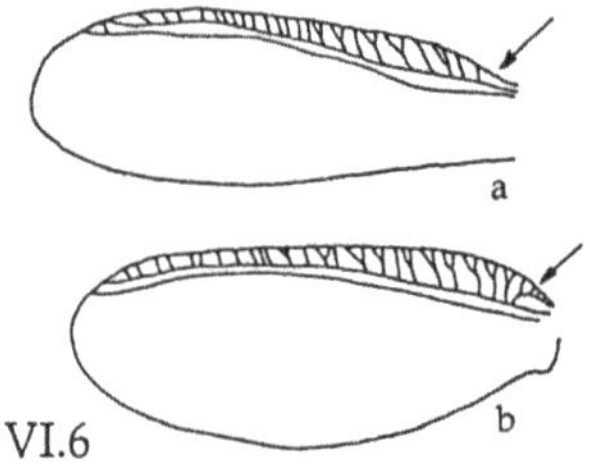

VI.6

2 Forewing without recurrent (turning back on itself) humeral vein at the base of the wing (VI.6a); small species 11 – 15mm long with narrow elongate wings *Micromus variegatus*

– Forewing with a recurrent humeral vein at the wing base (VI.6b) 3

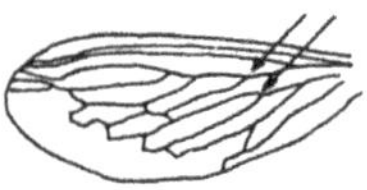

VI.7

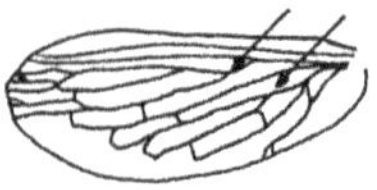

VI.8

3 Hindwing with first fork of the median vein as far out as first fork in the radial vein (VI.7); (forewings darkish and covered with brownish spots) *Kimminsia subnebulosa*

– Hindwing with first fork of the median vein well beyond the first fork in the radial vein (VI.8) 4

VI.9

4 Forewing with cross vein connecting vein Sc with R pale (VI.9); forewings pale yellow and hind border markings pale *Hemerobius lutescens*

– Forewing with cross vein connecting Sc with R dark; forewings brownish-yellow *Hemerobius humulinus*

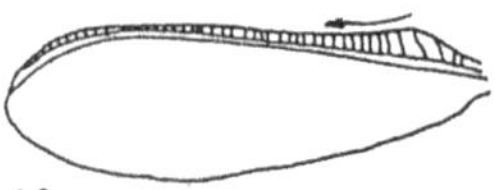

VI.10

5 Head, thorax and abdomen with strong black markings; 2nd segment of antennae markedly black *Chrysopa perla*

– Body without or with a few strong black markings; 2nd segment of antennae the same colour as the rest of the antennae 6

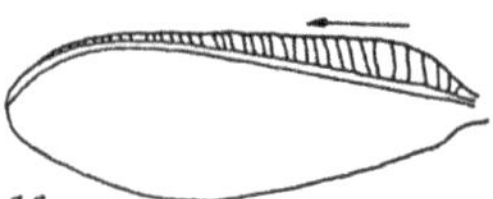

VI.11

6 Costal space of forewing narrowing abruptly after base (VI.10); large species 30 – 50mm *Nineta flava*

– Costal space of forewing not narrowing abruptly (VI.11), most specimens <35mm, except that some specimens of *Nineta vittata* reach 45mm long 7

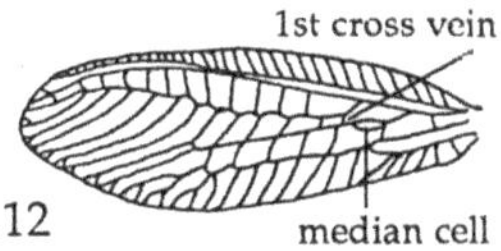

VI.12

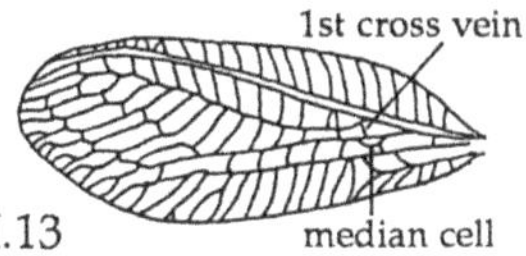

VI.13

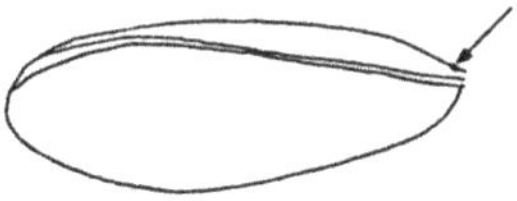

VI.14

7 Forewing with tip of median cell not extending as far as first cross vein (VI.12) *Chrysoperla carnea*

– Forewing with tip of median cell beyond 1st cross vein (VI.13) 8

8 Black spot on head between antennal bases 9

– This spot absent 10

9 Palps with black markings; black spot at base of costa in the forewing (VI.14) *Anisochrysa ventralis*

– Palps pale or with light brown markings; no black costal spot *Chrysopa septempunctata*

10 1st segment of antennae twice as long as broad; body rich green in life, fading to pale yellow after death *Nineta vittata*

– 1st antennal segment as long as broad; body whitish-green *Chrysotropia ciliata*

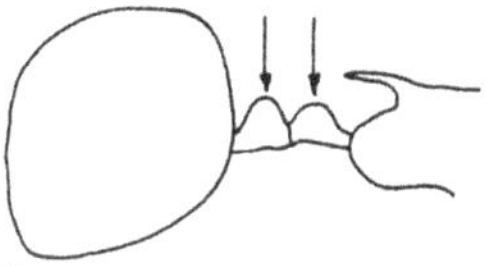

VII.1

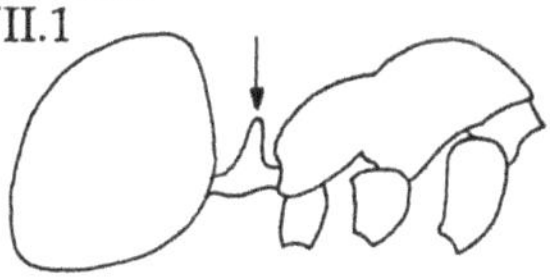

VII.2

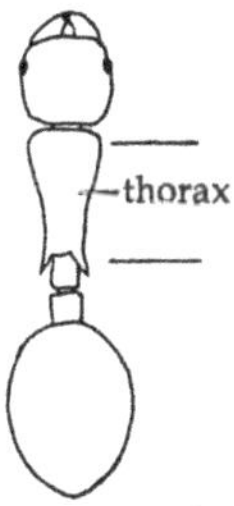

VII.3

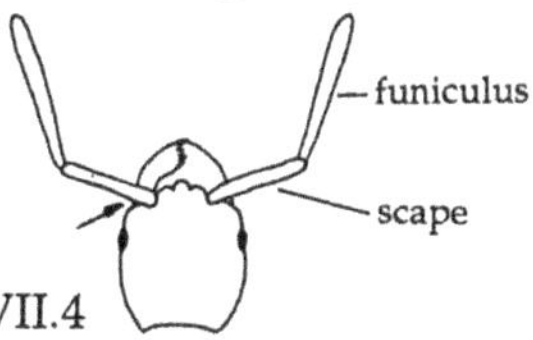

VII.4

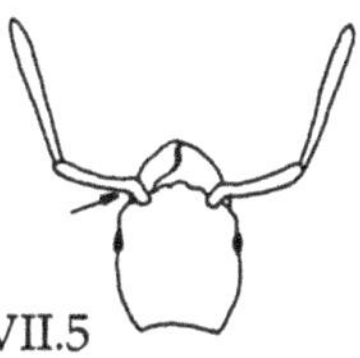

VII.5

Key VII
Ants, Formicidae

This key is based on worker ants. Usually male and queen ants are not seen attending aphid colonies.

1 Waist consists of 2 segments (VII.1) Myrmicinae 2

– Waist consists of 1 segment (VII.2) Formicinae 6

2 Reddish-brown ants with a pair of backwardly-directed projections on the rear of the thorax (VII.3); front of the thorax with rounded corners (VII.3); the last 3 antennal segments shorter than the rest of the funiculus (VII.4) *Myrmica* 3

– Not like this other Myrmicine genera (see Bolton & Collingwood, 1975; Brian, 1977)

3 Antennal scape without a right-angle bend near the base (VII.4); head shining 4

– Antennal scape bent at right angles near the base (VII.5); head matt 5

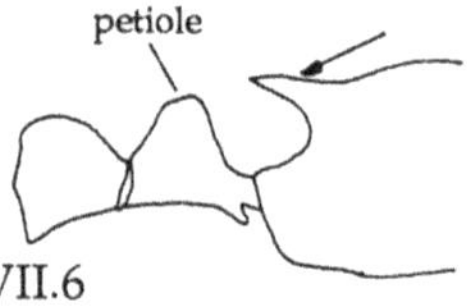

VII.6

4 Petiole with a rounded, almost pointed upper surface when viewed from the side (VII.6); thoracic spines short (VII.6); *Myrmica rubra* (L.)
(one of the commonest aphid-attending ants)

– Petiole with a flattened more convex upper surface when viewed from the side (VII.7); thoracic spines long (VII.7) *Myrmica ruginodis* Nylander

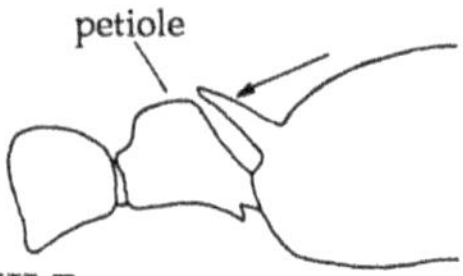

VII.7

5 Antennal scape with a large knob at the bend (VII.8) *Myrmica sabuleti* Meinert

– Antennal scape without a large knob at the bend; a small knob sometimes present *Myrmica scabrinodis* Nylander

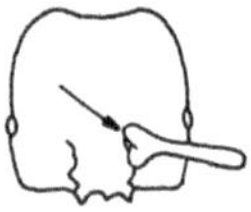
VII.8

6 Ocelli on top of the head large (VII.9); the spiracle on the side of the rear of the thorax is slit-shaped (VII.10) *Formica* (Wood ants) 7

– Ocelli on top of the head small (VII.11); the spiracle on the side of the rear of the thorax is rounded (VII.12) *Lasius* 10

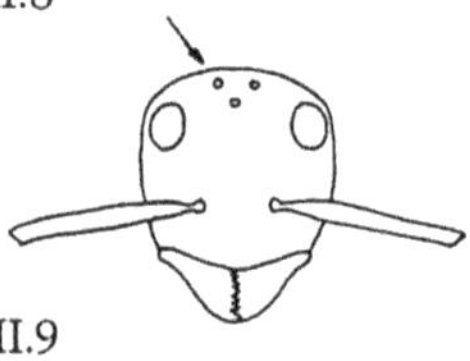
VII.9

VII.10

7 Thorax reddish-brown; ants nest in large mounds 8

– Thorax black 9

8 Eyes with small hairs between the facets; top of the head with long hairs when the face is viewed from the front *Formica lugubris* Zetterstedt

– Eyes and top of head without hairs *Formica rufa* (L.)
(a common southern species)

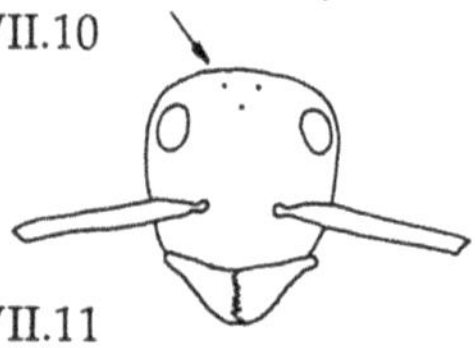
VII.11

VII.12

9 Undersides of the middle and hind femora with hairs (VII.13) *Formica lemani* Bondroit
(usually in upland areas)

– Undersides of the middle and hind femora without hairs *Formica fusca* (L.)
(usually a lowland species)

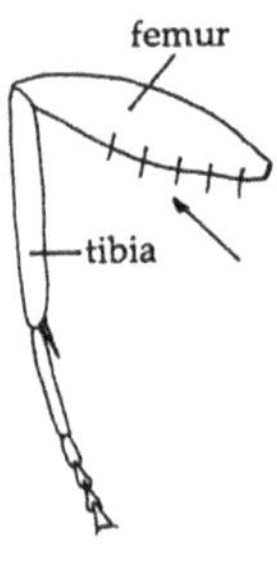

VII.13

10 Body dark brown to black; tibiae with long hairs amongst the short ones; body hairy and matt *Lasius niger* (L.)
(one of the commonest ants, making its nest under stones and in the soil, common in gardens)

– Not like this 11

11 Body yellow; tibiae with no long hairs amongst the short ones; long, erect hairs on top of the abdomen (these hairs > half the maximum width of the hind tibiae); nest consists of soil mound in grassland *Lasius flavus* (F.)

– Not like this other *Lasius* (see Bolton & Collingwood, 1975; Brian, 1977)

Guide to Carabidae and Staphylinidae (Ground and Rove beetles)

Carabidae

As outlined in chapter 2, many ground and rove beetles are potentially important aphid predators. Like earwigs they are mostly nocturnal predators. However, only in fields and in cereal crops have they been investigated to any extent (see table 8). Based on the results of these studies, descriptions are given below of those species in which aphids form a relatively major item in the diet. For other species and confirmation of those below use Forsythe (1987) or Lindroth (1974) to identify carabids and Joy (1932) for staphylinids.

B.1

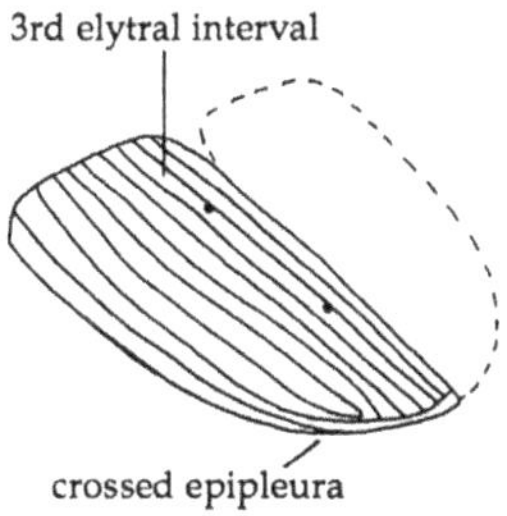

B.2

Bembidion lampros (Herbst)

Carabids of this genus are recognizable by their small size (< 7mm long; *B. lampros* is between 3 and 4.5mm long), by their symmetrical pattern of hairs and pits on both sides of the elytra (wing cases; hard, dark forewings) and by the outer pair of palps which have a club-like second segment and a small terminal segment (B.1). *B. lampros* is a very shiny brassy-blackish species with reddish legs and a smooth pronotum.

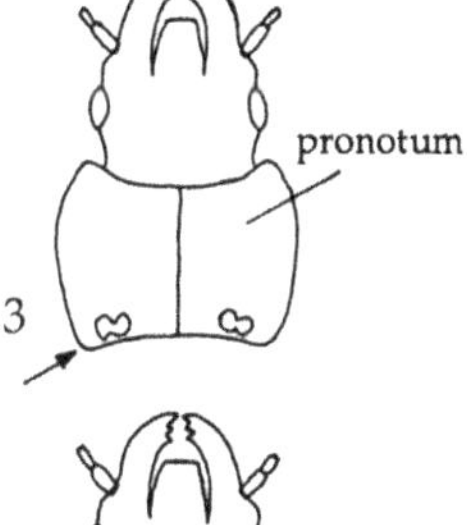

B.3

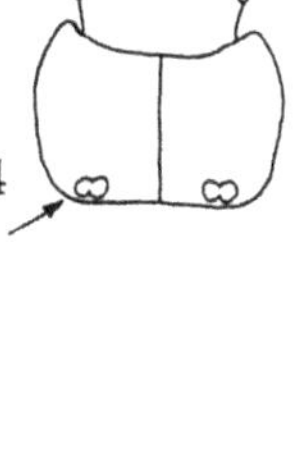
B.4

Pterostichus melanarius (Illiger)

This is a shiny black beetle between 12 and 18mm long with crossed epipleura (B.2). The third elytral interval has two pits (B.2) and the pronotum has, on each side, a double fovea (a dent in the cuticle) (B.3). The tibiae of the front legs are thickened. It can be distinguished from *Pterostichus madidus* (F.), the other species in the genus which is known to feed on aphids (see table 8), by the shape of the hind corner of the pronotum. In *P. madidus* this is rounded (B.4), but it is pointed in *P. melanarius* (B.3).

Table 8. *Beetle species recorded from fields and cereal crops with aphid remains in their digestive systems*

Beetle species	Notes
Carabidae	
Nebria brevicollis (F.)	abundant in many habitats; the larva is soil-burrowing
Notiophilus biguttatus (F.)	common, particularly in shady, dry sites
Loricera pilicornis (F.)	common in moist, shaded sites
Asaphidion flavipes (L.)	frequent in moist, open sites; more common in the east; few records from Scotland
Bembidion lampros (Herbst)	common during spring and early summer in many habitats
Pterostichus madidus (F.)	common in many habitats
Pterostichus melanarius (Illiger)	common in open sites
Calathus fuscipes Goeze	common in meadows and grassland
Synuchus nivalis Panzer	frequent in open sites
Agonum dorsale (Pontopiddan)	abundant, especially in open, dry habitats
Amara aenea (Degeer)	common in open sites
Amara familiaris (Duftschmid)	common in open sites
Amara plebeja (Gyllenhal)	common in grasslands on clay soils
Harpalus rufipes (Degeer)	abundant in open sites
Harpalus aeneus F.(=*affinis* (Schrank))	common in open sites
Demetrias atricapillus (L.)	abundant in open sites; no records north of Yorkshire
Staphylinidae	
Tachyporus chrysomelinus (L.)	very common everywhere
Tachyporus hypnorum (F.)	common
Tachyporus obtusus (L.)	common

Agonum dorsale (Pontoppidan)

This is an easily recognized beetle. It is between 6 and 8.5mm long with a coppery-green head and pronotum and reddish legs. The elytra are reddish except for a large, dark-greenish blotch in the rear half of the elytra. The epipleura are uncrossed and the mentum (the 'chin' – view from beneath) has a single tooth. This species is most common in spring and autumn.

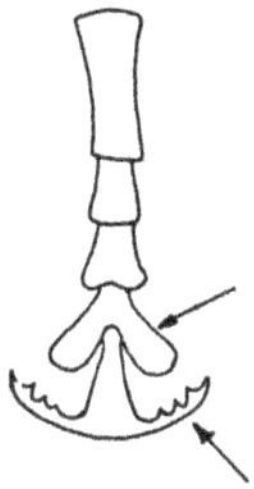

B.5

Demetrias atricapillus (L.)

This is a small (4 – 6mm long), pale yellowish beetle. It may be recognized easily by the two-lobed fourth segment of the tarsi and the three teeth on each claw (B.5). This species is unrecorded north of Yorkshire.

Staphylinidae

Tachyporus species

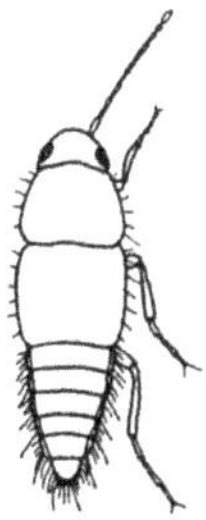

B.6

These are small colourful beetles 3 – 4mm long with a characteristic smooth, oval outline which tapers towards the rear (B.6). Along the side margins of the pronotum, elytra and abdomen are long, dark bristles. The sides of the abdomen are pressed flat along their edges. These are very active beetles often flexing the tip of the abdomen from side to side. The three common *Tachyporus* species may be identified provisionally using the key below, which relies on colour characters and should separate the majority of specimens. However, colour intermediates and doubtful specimens, possibly belonging to other *Tachyporus* species, need to be checked by an expert.

1 Abdomen black with reddish hind margins to the segments; the elytra red and the head and thorax black *Tachyporus hypnorum* (F.)

– Abdomen reddish with a dark tip 2

2 Elytra divided into a black half and a red half *Tachyporus obtusus* (L.)

– Elytra reddish, sometimes with black side margins *Tachyporus chrysomelinus* (L.)

5 Techniques

Handling aphids and predators in the laboratory

In chapter 1 techniques were described for collecting aphids and predators from the field. Before starting investigations in the laboratory it is important to be familiar with handling techniques so that damage and disturbance can be kept to a minimum.

Aphids and their predators are delicate and require gentle handling. Eggs and young larvae are particularly vulnerable. Shaking or sharply hitting the plant will loosen some aphids and predators but this is not a good technique in the laboratory. Gently blown carbon dioxide or tobacco fumes will 'knock down' aphids but the insects need long recovery periods. Knockdown methods for predators are not worthwhile. Individual aphids and predators can be transferred with a lightly moistened paintbrush, or in the case of large predators, with forceps. However, the safest method of transferring predators and aphid colonies is to leave them on their plants and move the whole plant or cut off the parts containing them. Eggs are usually stuck to the plant and have to be cut out.

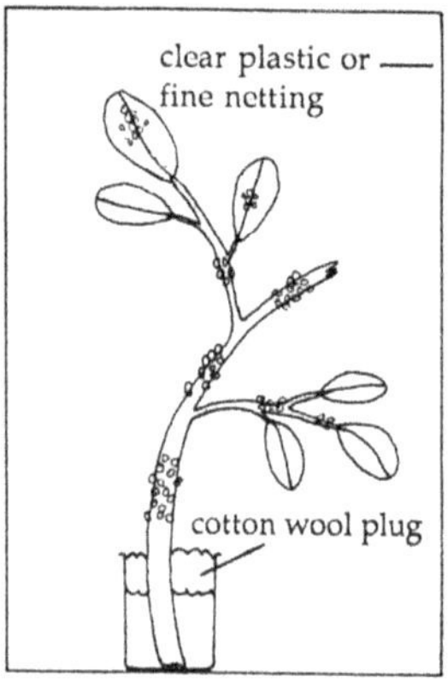

Fig. 18. Culturing aphids on a cut stem or small branch.

Rearing aphids

Having collected aphids from the field they need to be kept alive to provide food for predators. Aphids can be kept going on leaves or stems in plastic bags, sandwich boxes and similar containers for a day or so. They will keep for longer if the ends of the stems and small branches are kept in tubes of water. Use a cotton wool pad around the stem to keep it in place. This also helps to prevent aphids from drowning (fig. 18).

To obtain large numbers of aphids in continuous supply, pot up a few plants and grow aphid colonies in a greenhouse or rearing cage. If the plant begins to wilt under pressure from the aphids, transfer some of them to a new plant by detaching an infested leaf and placing it on a fresh plant. In a few hours the aphids should have walked to the new plant and begun to feed. Make sure a supply of fresh, young plants is always available.

In the field, whole plants, stems or branches can be 'sleeved' with netting (fig. 19). Root-dwelling aphids present extra difficulties. Methods devised for these special cases are described by Eastop & van Emden (1972). With artificial lighting and temperature control aphids can be kept going

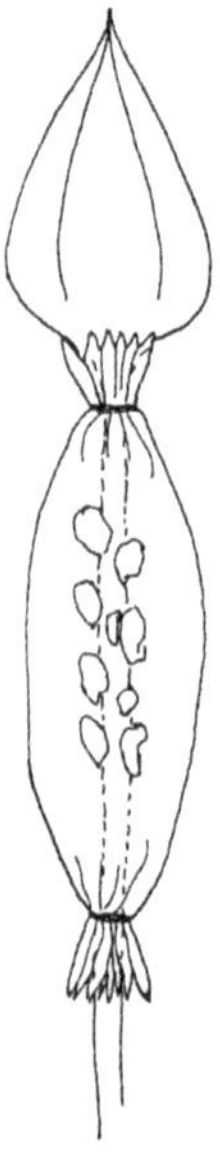

Fig. 19. 'Sleeving' an aphid infested rose stem.

throughout the winter (e.g. *Aphis fabae* on beans). Some aphid cultures can be started in the winter. For instance, small, overwintering nettle plants can be collected and potted up in a warm greenhouse. If aphid eggs are present they will hatch within a week or so.

Culturing aphid predators

Short-term cultures which last a few weeks can be easily maintained. Gravid (pregnant) female predators, caught in the field, are put in a container such as a plastic sandwich box with a piece of aphid-infested plant. Usually under these circumstances, dozens of eggs are produced within a few hours. Keep the females alive in separate tubes on a mixture of dilute honey and pollen, so that they can be reintroduced to aphids later and a regular supply of eggs obtained. By careful timing, it is possible to have supplies of each larval stage available at the same time.

As soon as the eggs hatch, larvae in groups of 5 to 10 should be transferred gently to a supply of aphids using a lightly moistened paintbrush. Alternatively, keep them separately in tubes or Petri dishes. A daily change of aphids is usually necessary. Try to keep larvae on one species of aphid in case a change of diet affects your experiments giving misleading results. Keep the various stages and species separate. Label every container. Some organization is obviously necessary when large numbers are involved. Don't try to culture everything you collect from the field. Plan investigations and keep only enough material for these. Otherwise it is very easy to spend all the time looking after the cultures! Expect some mortality in these cultures. Even in the best of circumstances some larvae die. You will be doing well if only 10 – 15% die before completing development.

As soon as larvae begin to pupate, separate them from the culture concerned and keep them individually in moist (but not wet) conditions until adults emerge. Hoverflies, lacewings and ladybirds can be wrapped up in damp tissue paper and put in tubes. Store the tubes in a cool place and check daily for any insects which have emerged. Aphid midges are best put on a layer of moist, sterile sand in a tube. Sterilize the sand first, by heating it in an oven for a few hours. Those larvae which are refusing to eat any more aphids, but have not pupated, are either aestivating (a temporary pause in development) or hibernating. Such larvae tend to be sluggish in their movements. Treat these individuals in the same way as normal puparia (as above). When winter begins, put any remaining larvae and puparia outdoors until the spring. This is because prolonged exposure to cold is often necessary to restart their development. Remember to label everything.

When the insects emerge, allow them plenty of time to develop fully their colour patterns before preserving them (see below). Usually this means allowing them to die naturally in the emergence tubes. If freshly-emerged adults are to be kept alive, feed them immediately on a dilute honey and pollen mix, as described above. Supply flower bugs and ladybirds with aphids for food.

If parasitoids emerge the following information should be recorded: a) date and place host collected, b) developmental stage of the host when collected, c) name of the host, aphid and plant (if in any doubt, place a ? before names), d) date the host pupated, e) date the parasitoid(s) emerged, f) whether the host was reared indoors or outdoors. Keep any remains, such as the host puparium, with the parasitoid (see 'Preserving' below).

Gravid hoverflies can be recognised by their swollen abdomens and the white colouring of the egg masses visible through the sides of the abdomen. They can be collected as they hover around aphid colonies. Gravid female aphid midges and lacewings can be collected from aphid colonies at night, unless there is a harsh wind when they tend not to fly. Flower bugs and ladybirds can be collected, reared and mated, and cultures started from eggs in the same way. As a further source of material, ladybird and lacewing egg batches can be collected from underneath leaves in the vicinity of aphid colonies.

It is difficult to maintain long-term cultures of some predators, such as hoverflies and lacewings, because it is not easy to get them to mate in the laboratory. It is simpler to utilize short-term cultures started from gravid females collected in the field. Nonetheless it would be of considerable interest to investigate techniques for obtaining laboratory mating in these groups as a separate project. Various avenues of investigation seem worthwhile pursuing. Try putting young male and female insects together in a tube and exposing them to very high light levels. Hand-mating may be tried. Hold a male and female by the wings, one in each hand, and bring them close together.

Rearing field-collected predators

The Petri dish-with-daily-change-of-aphids method is also a good way to rear field-collected predators. Try to maintain them on the aphid species from which they were collected, unless you know they can survive on another species. If you have to use field-collected predators for experiments, check whether they are parasitised or not by rearing them through at the end (chapter 1 discusses parasitoids of aphid predators). The behaviour of parasitised predators may be altered by the effects of parasitism, but little is known. This is a topic ripe for study.

Setting up experiments

The Petri dish-with-daily-change-of-aphids method, outlined above, is a standard technique used to investigate predator performance on several prey species (see section 3.1). Allow one predator per Petri dish, otherwise cannibalism or 'interference' may affect the results (section 3.2). Make sure the predators have enough prey by providing an excess of aphids. If necessary, determine how many aphids to give by monitoring how fast an individual predator consumes 25 aphids.

For some experiments, starved predators are needed. To starve a predator, first present it with aphids on a moistened brush and continue until aphids are consistently refused by the predator (use small aphids for small predators), then starve these satiated individuals for the desired period. Remember that young larvae cannot withstand much more than 24 hours starvation without becoming fatally weak. Predators which have been starved for prolonged periods sometimes recover if given water to drink. Offer them water by placing them in a drop or two.

There are a few simple precautions to take when recording behaviour. Avoid overly disturbing the predators when transferring them into an experimental arena (see 'Handling aphids and predators' above). Disturbed predators will try to escape rather than feed or search for prey. Signs of disturbance are animals which move to the top or sides of the cage or hide underneath leaves and remain still. Ground and rove beetles and earwigs are most prone to disturbance. Larval stages are much less affected. Replace overly disturbed predators with fresh individuals. If the predators are too active, they can be slowed down by chilling them in a refrigerator for a few minutes. Allow them to recover in the experimental arena. On the other hand, sluggish predators may need a gentle prod to get them moving. Inactivity is, however, a normal part of the behaviour of many predators. They will often rest immediately after finishing an aphid. When moving about they can stop suddenly for no obvious reason. Activity is an important means whereby bird predators find prey and resting in aphid predators may help to conceal them from these avian enemies.

Do not use the same individual predator repeatedly for a number of tests – its behaviour may change as a result of experience. Instead, repeat each experiment several times using fresh individuals each time. This will give an idea of the range of variation and is necessary for subsequent statistical analyses (see below). Choose experimental subjects from a group fed and reared in the same way so that any variation arises from your experimental manipulation and not from accidental differences in age or history between the insects tested. Finally, have patience. Be prepared to watch your predators for two or three hours at a time. You will be doing well to manage a few replicates each day. Often the best progress is made slowly.

How to present the results of research

Writing up is an important part of a research project, particularly when the findings are to be communicated to others. A really thorough, critical investigation which has established new information of general interest may be worth publishing if the animals on which it is based have been identified with certainty. Journals which publish short papers on insect biology include the *Entomologists' Monthly Magazine, Entomologist's Gazette, Bulletin of the Amateur Entomologists' Society* and the *Journal of Biological Education*. Those unfamiliar with publishing conventions are advised to examine current issues of the journals to see the type of articles which are published, and then to write a paper along similar lines, keeping it as short as is consistent with the presentation of enough information to establish conclusions. It is then time to consult an appropriate expert who can give advice on whether and in what form the material might be published.

It is an unbreakable convention of scientific publication that results are reported with scrupulous honesty. It is essential, therefore, to keep detailed and accurate records throughout the investigation, and to distinguish in the write-up between certainty and probability, and deduction and speculation. In many cases it will be necessary to apply appropriate statistical techniques to test the significance of the findings. A book such as *The OU Project Guide* (Chalmers & Parker, 1985) or Parker's *Introductory Statistics for Biology* (1973) will help, but this is an area where expert advice can contribute much to the planning, as well as to the analysis, of the work.

Preserving

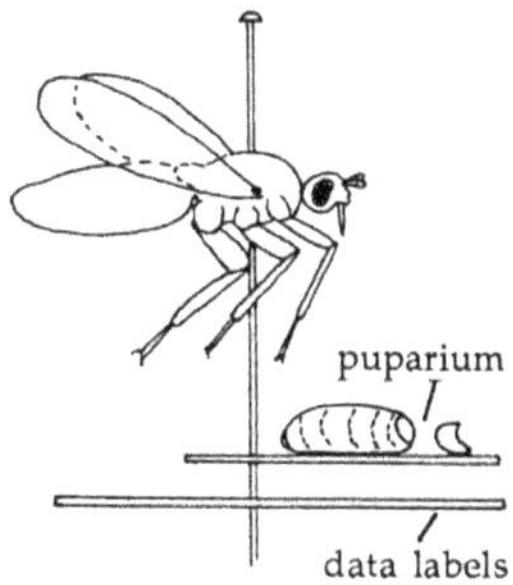

Fig. 20. Direct pinning an insect.

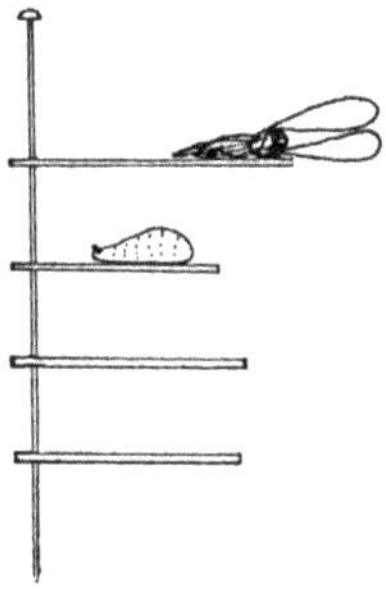

Fig. 21. 'Staging' an adult insect.

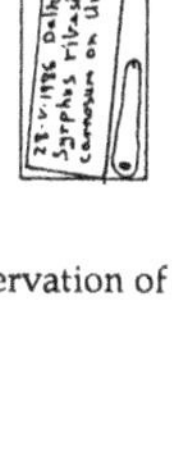

Fig. 22. Preservation of larva in spirit.

When you have finished an investigation, devote time to the preparation of your specimens for long-term preservation. This is part of good scientific practice. You or other workers may need to study the material again, especially if you have published any information about it. If you have completely finished with your studies and no longer wish to keep your collection, consider offering it to a natural history museum. This is the best way of ensuring the future availability of your material for other workers, especially if you mention in your publications which museum has accepted the material. Take scrupulous care to label it accurately.

Adult insects can be left in their tubes to die or can be killed with strips of filter or tissue paper lightly moistened with ethyl acetate. Alternatively, leave your specimens in a freezer for a few hours. This is also a good method of temporary storage (up to 6 months). Unless you intend to keep the insects in a freezer, preserve them within a few hours of death, otherwise they will become dry and brittle.

Adults can be pinned through the thorax (fig. 20). Smaller insects, such as flower bugs and ladybirds, are glued to pieces of card using water-soluble gum (fig. 21). The card is then pinned. Data labels are attached to the same pin beneath. Glue any remains from the pupal stage to a piece of card and pin that also. Keep the specimens in air-tight boxes which are obtainable from dealers, or make your own from sandwich boxes floored with expanded polyethylene foam. You may be able to borrow suitable boxes from a museum if you promise to donate the specimens. Small or soft-bodied insects such as aphids, aphid midges and larvae are best preserved in small plastic-capped vials containing 70% alcohol. Include data labels written in pencil on thick card (fig. 22). Live aphids can be dropped into alcohol without further preparation, but predator larvae should first be placed in hot water (60°C) for a few seconds. This not only fixes them (keeps them stable) but also has the advantage of causing their segments to extend, which leaves all of their important features visible.

Within a few weeks of preservation most aphids and predator larvae will lose their colour and turn pale. It is, therefore, wise to record their colours when alive, by photography, drawing and note-taking.

Further reading

Finding books

Some of the books and journals listed here will be unavailable in local and school libraries. It is possible to make arrangements to see or to borrow such works by seeking permission to visit the library of a local university, or by asking your local public library to borrow the work (or a photocopy of it) for you via the British Library, Document Supply Centre. This may take several weeks, and it is important to present your librarian with a reference which is correct in every detail. References are acceptable in the form and order given here, namely the author's name and date of publication, followed by (for a book) the title and publisher or (for a journal article) the title of the article, the journal title, the volume number and the first and last pages of the article.

The Handbooks for the Identification of British Insects, published by the Royal Entomological Society of London, may be bought from the British Museum (Natural History), Cromwell Road, London SW7 5BD.

References

Alford, D.V. (1984). *A Colour Atlas of Fruit Pests: their Recognition, Biology and Control.* London. Wolfe Publishing Ltd.

Anderson, N.H. (1962). Bionomics of six species of *Anthocoris* (Heteroptera: Anthocoridae) in England. *Transactions of the Royal Entomological Society of London* **114**, 67-95.

Banks, C.J. (1957). The behaviour of individual coccinellid larvae on plants. *British Journal of Animal Behaviour* **5**, 12-24.

Bansch, R. (1966). On prey-seeking behaviour of aphidophagous insects. In: Hodek, I., ed. *Ecology of Aphidophagous Insects*: 123-128. The Hague: Junk Publishers.

Bhatia, M.L. (1939). Biology, morphology and anatomy of aphidophagous syrphid larvae. *Parasitology* **31**, 78-129.

Blackman, R.L. (1967). The effects of different aphid foods on *Adalia bipunctata* L. and *Coccinella 7-punctata* L. *Annals of Applied Biology* **59**, 207-219.

Blackman, R.L. (1974). *Aphids*. London: Ginn.

Bolton, B. & Collingwood, C.A. (1975). Hymenoptera, Formicidae. *Handbooks for the Identification of British Insects* **6**, no. 3c.

Brian, M.V. (1977). *Ants*. New Naturalist, no. 59. London: Collins.

Canard, M., Semeria, Y. & New, T.R. (1984). *Biology of Chrysopidae*. Series Entomologica 27. The Hague: Junk Publishers.

Carter, C.I. (1971). *Conifer Woolly Aphids (Adelgidae) in Britain*. Forestry Commission Bulletin No. 42, London.

Carter, M.C., Sutherland, D. & Dixon, A.F.G. (1984). Plant structure and the searching efficiency of coccinellid larvae. *Oecologia* **63**, 394-397.

Chalmers, N. & Parker, P. (1985). *The OU Project Guide*. London: Field Studies Council.

Chandler, A.E.F. (1968a). Height preferences for oviposition of aphidophagous Syrphidae (Diptera). *Entomophaga* **13**, 187-195.

Chandler, A.E.F. (1968b). A preliminary key to the eggs of some of the commoner aphidophagous Syrphidae (Diptera) occurring in Britain. *Transactions of the Royal Entomological Society of London* **120**, 199-217.

Chandler, A.E.F. (1968c). Some host-plant factors affecting oviposition by aphidophagous Syrphidae (Diptera). *Annals of Applied Biology* **61**, 415-423.

Chandler, A.E.F. (1969). Locomotory behaviour of first instar larvae of aphidophagous Syrphidae (Diptera) after contact with aphids. *Animal Behaviour* **17**, 673-678.

Chinery, M. (1976). *A Field Guide to the Insects of Britain and Northern Europe*, 2nd edition. London: Collins.

Collin, J.E. (1961). *British Flies* 6. *Empididae*. Cambridge: Cambridge University Press.

Colyer, C.N. & Hammond, C.O. (1968). *Flies of the British Isles*, 2nd edition. London: Warne.

Dean, G.J. (1983). Survival of some aphid (Hemiptera: Aphididae) predators with special reference to their parasites in England. *Bulletin of Entomological Research* **73**, 469-480.

Disney, R.H.L. (1983). Scuttle Flies (Diptera, Phoridae). *Handbooks for the Identification of British Insects* **10**, no.6.

Dixon, A.F.G. (1959). An experimental study of the searching behaviour of the predatory coccinellid beetle *Adalia decempunctata* (L.). *Journal of Animal Ecology* **28**, 259-281.

Dixon, A.F.G., Martin-Smith, M. & Subramanian, G. (1965). Constituents of *Megoura viciae* Buckton. *Journal of the Chemical Society*: 1562-1564.

Dixon, T.J. (1960). Key to and descriptions of the third instar larvae of some species of Syrphidae (Diptera) occurring in Britain. *Transactions of the Royal Entomological Society of London* **112**, 345-379.

Dunn, J.A. (1949). The parasites and predators of potato aphids. *Bulletin of Entomological Research* **40**, 97-122.

Eastop, V.F. & van Emden, H.F. (1972). *Aphid Technology*. London: Academic Press.

Edwards, C.A., Sunderland, K.D. & George, K.S. (1979). Studies on polyphagous predators of cereal aphids. *Journal of Applied Ecology* **16**, 811-823.

Edwards, R. (1980). *Social Wasps, their Biology and Control*. East Grinstead: Rentokil Ltd.

El-Ziady, S. & Kennedy, J.S. (1956). Beneficial effects of the common garden ant, *Lasius niger* L. on the black bean aphid, *Aphid fabae* Scopoli. *Proceedings of the Royal Entomological Society of London* (A) **31**, 61-65.

Evans, A.F. (1976a). The role of predator size ratio in determining the efficiency of capture by *Anthocoris nemorum* and the escape reactions of the prey, *Acyrthosiphum pisum*. *Ecological Entomology* **1**, 85-90.

Evans, A.F. (1976b). The searching behaviour of *Anthocoris confusus* (Reuter) in relation to prey density and plant surface topography. *Ecological Entomology* **1**, 163-169.

Evans, H.F. (1976c). Mutual interference between predatory anthocorids. *Ecological Entomology* **1**, 283-286.

Fitton, M.G. & Rotheray, G.E. (1982). A key to the European genera of diplazontine ichneumon-flies, with notes on the British fauna. *Systematic Entomology* **7**, 311-320.

Fonseca, E.C.M.d'Assis (1978). Diptera Orthorrhapha Brachycera Dolichopodidae. *Handbooks for the Identification of British Insects* **9**, no.5.

Forsythe, T.G. (1987). *Common Ground Beetles*. Naturalists' Handbooks 8. Slough: Richmond Publishing Co. Ltd.

Fraser, F.C. (1959). Mecoptera Megaloptera Neuroptera. *Handbooks for the Identification of British Insects* **1**, nos. 12 & 13.

Gilbert, F.S. (1986). *Hoverflies*. Naturalists' Handbooks 5. Cambridge: Cambridge University Press.

Harris, K.M. (1966). Gall midge genera of economic importance (Diptera: Cecidomyiidae. 1. Introduction and subfamily Cecidomyiinae; supertribe Cecidomyiidi. *Transactions of the Royal Entomological Society of London* **118**, 313-358.

Harris, K.M. (1973). Aphidophagous Cecidomyiidae (Diptera): Taxonomy, biology and assessment of field populations. *Bulletin of Entomological Research* **63**, 305-325.

Hassell, M.P. (1976). *The Dynamics of Competition and Predation*. London: Arnold.

Hodek, I. (1973). *Biology of Coccinellidae*. The Hague: Junk Publishers.

Jones, D. (1983). *Spiders of Britain and Northern Europe*. Feltham: Hamlyn Books.

Joy, N.H. (1932). *A Pactical Handbook of British Beetles* (2 vols.). London: Witherby. (Reprinted in 1976 by E.W. Classey Ltd.).

Killington, F.J. (1936). *A Monograph of the British Neuroptera* 1 & 2. London: Ray Society. (Vol. 2 published in 1937).

Lindroth, C.H. (1974). Coleoptera, Carabidae. *Handbooks for the Identification of British Insects* 4, no.2.

Linssen, E.F. (1959). *Beetles of the British Isles* (2 vols). London: Warne.

Luff, M.L. (1978). Diel activity patterns of some field Carabidae. *Ecological Entomology* **3**, 53-62.

McClintock, D. & Fitter, R.S.R. (1982). *Collins Pocket Guide to Wild Flowers*. London: Collins.

Majerus, M.E.N. & Kearns, P.W.E. (1989). *Ladybirds*. Naturalists' Handbooks 10. Slough: Richmond Publishing Co. Ltd.

New, T.R. (1975). The biology of Chrysopidae and Hemerobiidae (Neuroptera) with reference to their usage as biocontrol agents: a review. *Transactions of the Royal Entomological Society of London* **127**, 115-140.

Parker, J.R. (1918). The life history and habits of *Chloropisca glabra* Meig., a predaceous Oscinid (Chloropid). *Journal of Economic Entomology* **11**, 368-380.

Parker, R.E. (1973). *Introductory Statistics for Biologists*. London: Edward Arnold.

Phillips, R. (1978). *Wild Flowers of Britain*. London: Pan.

Pollard, E. (1971). Hedges. VI. Habitat diversity and crop pests: a study of *Brevicoryne brassicae* and its syrphid predators. *Journal of Applied Ecology* **8**, 751-780.

Pontin, A.J. (1958). A preliminary note on the eating of aphids by ants of the genus *Lasius*. *Entomologists' Monthly Magazine* **94**, 9-11.

Pope, R.D. (1953). Coleoptera, Coccinellidae & Sphindidae. *Handbooks for the Identification of British Insects* **5**, no.7.

Roberts, M.J. (1971). On the locomotion of Cyclorrhapha maggots (Diptera). *Journal of Natural History* **5**, 583-590.

Rotheray, G.E. (1979). The biology and host searching behaviour of a cynipoid parasite of aphidophagous syrphid larvae. *Ecological Entomology* **4**, 79-82.

Rotheray, G.E. (1981). Host searching and oviposition behaviour of some parasitoids of aphidophagous Syrphidae. *Ecological Entomology* **6**, 79-87.

Rotheray, G.E. (1983). Feeding behaviour of *Syrphus ribesii* and *Melanostoma scalare* on *Aphis fabae*. *Entomologia experimentalis et applicata* **34**, 148-154.

Rotheray, G.E. (1984). Host relations, life cycles and multiparasitism in some parasitoids of aphidophagous Syrphidae (Diptera). *Ecological Entomology* **9**, 303-310.

Rotheray, G.E. (1986). Colour, shape and defence in aphidophagous syrphid larvae (Diptera). *Zoological Journal of the Linnean Society* **88**, 201-216.

Rotheray, G.E. & Martinat, P. (1984). Searching behaviour in relation to starvation of *Syrphus ribesii*. *Entomologia experimentalis et applicata* **36**, 17-21.

Russel, R.J. (1970). The effectiveness of *Anthocoris nemorum* and *A. confusus* (Hemiptera: Anthocoridae) as predators of the sycamore aphid, *Drepanosiphum platanoidis*. *Entomologia experimentalis et applicata* **13**, 194-207.

Sands, W.A. (1957). The immature stages of some British Anthocoridae (Hemiptera). *Transactions of the Royal Entomological Society of London* **109**, 295-310.

Sankey, J.H.P. & Savory, T.H. (1974). *Synopses of the British Fauna (New Series)* No. 4. *British Harvestmen*. The Linnean Society of London. London: Academic Press.

Scott, E.I. (1939). An account of the developmental stages of some aphidophagous Syrphidae (Diptera) and their parasites (Hymenoptera). *Annals of Applied Biology* **26**, 509-532.

Shah, M.A. (1982). The influence of plant surfaces on the searching behaviour of coccinellid larvae. *Entomologia experimentalis et applicata* **31**, 377-380.

Southwood, T.R.E. (1973). The insect/plant relationship–an evolutionary perspective. *Symposium of the Royal Entomological Society of London* **6**, 3-30.

Southwood, T.R.E. & Leston, D. (1959). *Land and Water Bugs of the British Isles*. London: Warne.

Stroyan, H.L.G. (1984). Aphids – Pterocommatinae and Aphidinae (Aphidini). *Handbooks for the Identification of British Insects* **2**, no.6.

Stubbs, A.E. & Falk, S.J. (1983). *British Hoverflies: an Illustrated Identification Guide*. London: British Entomological and Natural History Society.

Sunderland, K.D. (1975). The diet of some predatory arthropods in cereal crops. *Journal of Applied Ecology* **12**, 507-515.

Unwin, D. (1981). A key to the families of British Diptera. *Field Studies* **5**, 513-533. (An AIDGAP key.)

Unwin, D. (1984). A key to the families of British Coleoptera (and Strepsiptera). *Field Studies* **6**, 149-197. (An AIDGAP key.)

van Emden, F.I. (1942). Larvae of British Beetles. III. Keys to the families. *Entomologist's Monthly Magazine* **92**, 206-272.

Varley, G.C., Gradwell, G.R. & Hassell, M.P. (1973). *Insect Population Ecology*. London: Blackwell Scientific Publications.

Vickerman, G.P. & Sunderland, K.D. (1975). Arthropods in cereal crops: nocturnal activity, vertical distribution and aphid predation. *Journal of Applied Ecology* **12**, 755-765.

Walker, M.F. (1961). Some observations on the biology of the ladybird parasite *Perilitus coccinellae* (Shrank) (Hym. Braconidae), with special reference to host selection and recognition. *Entomologist's Monthly Magazine* **97**, 240-244.

Way, M.J. (1954). Studies on the association of the ant *Oecophylla longinoda* with the scale insect *Saissetia zanzibarensis*. *Bulletin of Entomological Research* **45**, 114-154.

Way, M.J. (1963). Mutualism between ants and honeydew-producing Hemiptera. *Annual Review of Entomology* **8**, 307-344.

Yeo, P.F. & Corbet, S.A. (1983). *Solitary Wasps*. Naturalists' Handbooks 3. Cambridge: Cambridge University Press.

Useful addresses

Entomological equipment suppliers

Watkins and Doncaster, Four Throws, Hawkhurst, Kent

Worldwide Butterflies Ltd., Compton House, nr. Sherborne, Dorset DT9 4QN

Suppliers of entomological books, new and secondhand

E.W. Classey Ltd., P.O.Box 93, Faringdon, Oxon SN7 7DR

L. Christie, 129 Franciscan Road, Tooting, London SW17 8DZ

The Richmond Publishing Co. Ltd., P.O.Box 963, Slough SL2 3RS
(Supply AIDGAP keys, Naturalists' Handbooks and Ladybird colour identification chart)

Entomological and other societies

Amateur Entomologists' Society, 355 Hounslow Road, Hanworth, Feltham, Middlesex, TW13 5JH

British Entomological and Natural History Society, c/o The Alpine Club, 74 South Audley Street, London W1Y 5FF

Royal Entomological Society of London, 41 Queen's Gate, London SW7 5HV

Index

www.ingramcontent.com/pod-product-compliance
Ingram Content Group UK Ltd.
Pitfield, Milton Keynes, MK11 3LW, UK
UKHW051949271225
466430UK00001B/3

9 781784 275105